SDGs and Textiles

Editor-in-Chief

Hafeezullah Memon, College of Textile Science and Engineering, International Institute of Silk, Zhejiang Sci-Tech University, Hangzhou, China

The book series "SDGs and Textiles" addresses the strategies to achieve sustainable development goals (SDGs) in the present, past, and future. It presents books about the present and future policies of textile ministries of different countries, and books related to sustainability education around different parts of the world in the textile sector. Moreover, it would welcome the conference proceeding related to SDGs and Textiles. The series would cover books comparing the sustainability and SDGs of different institutions and countries. The individual book volumes in the series are thematic. The goal of each book is to give readers a comprehensive overview of a different area of sustainability in the textile sector. As a collection, the series provides valuable resources to a broad audience in academia, the research community, industry, and anyone looking to expand their knowledge of SDGs and Textiles.

Textiles and life are together – life cannot be separated from textiles as it is the most important need for human beings after food. In 2015, the United Nations General Assembly proposed 17 interlinked global goals to be achieved by 2030. Since then, academia and industry have paid much attention to achieving these goals. Textile found its close relation with almost all of these 17 goals.

SDG 1 - No Poverty: Poverty would never be overcome by a charity only; it is essential to develop people's skills to have a better and wealthy life. Thus, the textile can be considered an excellent discipline to achieve this goal by creating jobs and small and medium businesses.

SDG 2 - Zero Hunger: Through the effective utilization of advanced application of Agrotech Textiles, it is possible to have higher crop yields and save crops from rough weather, unexpected rains, floods, insects, etc.; thus, geotextiles play an essential in achieving this goal of sustainable development.

SDG 3 - Good Health & well-being: There has been much health consciousness after Covid19, and medical textiles assist in getting good health and well-being.

SDG 4 - Learning & Education: Textile or fashion has remained a significant discipline for societies for ages, and there has always remained much to explore in this field. Textile-related universities may play a vital role by offering free access to their education resources, training and spreading information among the locals.

SDG 5 - Gender Equality: The textile sector is one of the industrial sectors that accepted gender equality long ago; in particular, the garment sector has more females than males. Thus, the textile sector has been doing gender equality. Moreover, there has been a recent trend for Gender Neutral Clothing, which need worth studying and may further assist gender equality.

SDG 6 - Clean Water & Sanitation: Textiles could be achieved through filtration, and of course, textile is one of the critical materials for filtration.

SDG 7 - Affordable & Clean Energy: With the recent advancement in material science and engineering, the textile sector has come on the front for, not only by using this clean energy during textile production but also by assisting the production of this clean energy, either in the form of wind turbines blades made of textile composites or by energy harvesting from T-Shirts, etc.

SDG 8 - Decent Work: Recently, there has been much attention that the textile workers are not paid well, labor rights are not cared about, etc.

SDG 9 - Industry and innovation: Textile Industry always follows innovation; the textile companies that do not chase innovation cannot survive in the market.

SDG 10 - Reduced Inequalities: Getting better life and well-being would help reduce inequalities in the textile industry.

SDG 11 - Sustainable Cities: Sustainable Textile Cities through Buildtech and transport textiles.

SDG 12 - Consumption and Production: Textile and garment consumption and production all come under.

SDG 13 - Climate Action: Oekotech or Ecotech Textile, waste management of textiles are upfront to achieve this goal of sustainable development.

SDG 14 - Life Below Water: Mitigating microfiber waste in rivers and oceans may come under the context of it. There has been much attention on this subject after passing the bill at the parliament level of the UK.

SDG 15 - Life on Land: Geotech or Geotextiles studies life on land.

SDG 16 - Peace, Justice, and Strong Institutions: Protective textiles are doing their best to achieve peace, justice, and strong institutions.

SDG 17 - Partnerships for the Goals: The application of textiles to achieve sustainable development goals is only an example. In all textiles sectors, combined efforts of all the goals are essential to achieve true sustainability.

Gayathri Madubhani Ranathunga
Editor

Textiles of Sri Lanka

Sustainable Innovations and Practices

Springer

Editor
Gayathri Madubhani Ranathunga
Department of Textile and Apparel
Engineering
Fashion Design and Product Development
Degree Course
Faculty of Engineering
University of Moratuwa
Moratuwa, Sri Lanka

ISSN 2948-1236 ISSN 2948-1244 (electronic)
SDGs and Textiles
ISBN 978-981-97-0568-9 ISBN 978-981-97-0569-6 (eBook)
https://doi.org/10.1007/978-981-97-0569-6

This Springer imprint is published by the registered company Springer Nature Singapore Pte Ltd.
The registered company address is: 152 Beach Road, #21-01/04 Gateway East, Singapore 189721, Singapore

Paper in this product is recyclable.

Foreword

A rich collection of research on an industry in one country awaits you as the subject of this book. The book is all about Sri Lanka's textile and apparel industry explored along two roots—one with a very long history and the other less than 100 years. Both forms of industry are present across the country with different traditions and practices keeping the older form quite alive as well. As in textiles, the woven story of the narrative across 11 chapters links the industry in different ways to the Sustainable Development Goals—the all-consuming task of all the planetary inhabitants today. Some of us understand the dire consequences of missing these goals! Many of us I am afraid are not and research showcasing the progress being made or the absence is thus quite important.

Perhaps not many countries have their history starting with a storyline connecting to the textile industry and that is what Sri Lanka has. When King Vijaya and his group landed in Sri Lanka on or around 543 BC their first contact was a lady weaving spinning thread. Kuveni, the lady weaving spinning thread was a queen in Sri Lanka as per the Pali chronicle Mahavansa. She is from the indigenous inhabitants of Thambapani (then Sri Lanka). With this romantic occurrence—the legends vary in many ways but I select the joyful version—Vijaya married Kuveni and settled in Sri Lanka and the written history of Mahavansa then moves on.

Moving forward 2500 years the textile industry in Sri Lanka has become the sector to give a significant export earning to the country—a real GDP contributor. The industry over the years has only grown in importance in all its formats—economic, social, and environmental. Today it is the triple bottom-line considerations that we do in a well-meaning sustainable future-focused analysis to measure success of an industry, keeping in mind the future generations. The contributions as the research reveals have not been always positive as some growth has resulted in environmental and social stresses. One could foresee economic stresses as well as the industry's resources coming from the region with the majority of the consumers having to be reached with long supply chains. Primarily in the mass market apparel industry, it is the specialist production human capital the country provides. The compilation offers an analytical view of the industry from different directions. All chapters tackle those questions to different degrees keeping the UN goals of SDG as benchmark.

It is clear that while the industry sector is doing reasonably well many issues are exist. I hope that the findings do not reflect the Kuveni's curse that our history also alludes to. Authors are quite specific in their research recommendations too in ways to overcome the challenges and clear with details on the possibilities. That human capital sourced from Sri Lanka had served Bangladesh for instance quite well, and the industry there had surpassed Sri Lanka. Interesting to ponder–does that mean the Sri Lankan industrial environment is not conducive to dynamic growth though the technical abilities are excellent? Taking SDG 9 more seriously across the board the role of innovation in the Sri Lankan textile industry is enabling its competitiveness. The sufficiency of innovation process is opened up without an answer here. Though the exact details are not coming through the chapters Sri Lanka can boast of global firsts as well as perhaps the highest production floor areas per unit area in green apparel manufacturing. The "garments without guilt" projects a strong positive socially responsible image of the Sri Lankan Apparel industry.

Two chapters dwell into the handloom sector where one could witness history preserved. Interesting to read about the oral traditions keeping the handloom sector alive and how that is taking place in specific geographic locations. There was a significant impact on this sector when the economy opened up in 1977. However, the existence of it still in different parts of the country from the south to the central hills as described here, the research demonstrates the possible resilience when communities persist and society too responds positively. Yet there is a divide between this sector meeting local needs while the other serves globally and with a clear difference in economic value. Happening at different speeds SDGs are likely to be compromised.

The discussion also opens up the opportunity to bring natural fibres to create a sustainable manufacturing enterprise. The opportunity to scale up with natural fibres, i.e. banana as an example is presented though a technology gap exists in yielding required amounts of natural fibre. What is the impact of society embracing SDG 12 in reducing the demand burden? It is almost inconceivable to realise sustainable development in the absence of SDG 12. The modern apparel industry facing SDG 12 headwinds is still a slowly developing phenomenon as the current norm is more is needed by more. The handloom sector as a slow sector offers us a way to understand the value of a slow pace and controlling the demand.

The textile and apparel industry is well known for the significant no. of employment provided for females. The positivity that has come about by broad-basing the textile industry across Sri Lanka has come out. From the side of poverty, the industry has supported addressing multi-dimensional poverty. The glass ceiling appears to exist and definitely quite a significant amount of sticky floors. An interesting study with significant value is the dress choice of senior female executives and the correlation between the choice of dress and career growth. While reading this I visualised the picture in the Indian mission control room when the all-flight control personnel applauded when Chandrayan landed. Not exactly a comparison of apples with apples yet reading this I see there could be strong sociological considerations to be drawn from this study.

The circularity aspects of the industry have received attention. As good as new covers the second-hand garments. The upcycling has today come in, and some in the industry are on global stage. The industry has embraced management techniques such as ISO14062: 2002 to embed sustainability drivers into organisational strategy. These steps however are with the large mass market apparel manufacturers who cater to distant markets. There is a need to customise and communicate these findings to those who operate at lower rungs. If one scales up plant dyes and carries out mordant-free dyeing you are in cleaner production from the beginning. Then as discussed testing is on sustainability practices, and testing themselves can be quite different and environmentally benign. All these thoughts get generated as one reads through. Important when designing change.

Quite a collection of practices, possibilities, and ideas is what this collection represents. Can the Pearl of the Indian Ocean rediscover the industry with the collective wisdom presented here? I see only the missing part of eco-innovation discussed at some length which is the prerequisite for the transformation. The seeds have been planted surely. This comes out as a unique collection covering one industry carried out by examining the role of driving the industry to develop their aspirational basis parallel to SDGs to grow and become meaningful to the economy.

Ajith de Alwis
B.Sc. Eng (Moratuwa), MBA (PIM-SriJ),
Ph.D. (Cambridge) FNASSL
Senior Professor in Chemical and Process Engineering
University of Moratuwa
Sri Lanka

Chief Innovation Officer
National Innovation Agency
Sri Lanka

Preface

This book highlights the role of sustainable development in ancient practices and modern innovations in the textiles of Sri Lanka. Today, the textile industry of Sri Lanka has developed along two roots: the domestic handloom industry, which caters mainly to local consumers, and the apparel industry, which caters to global context needs. Both root causes have identified that the SDGs are aspirational to soar the country's economy.

The book reveals the textiles of Sri Lanka from the historical age to current tech-fabrics. The textile industry is one of the oldest in Sri Lanka, and it goes back to 2500 years of the country's history. The earliest handloom weaving was based on domestic weavers of Sri Lanka and practised as an indigenous craft. The most sought-after concepts and practices of maximum utilisation, zero waste, eco-friendly production process, reuse, upcycling, recycling, and longevity, such as sustainable development concepts and life, have been in the traditional society of Sri Lanka. The system was rooted in civil society; occupation and life became one. The employee and the employer were the craftsmen, and the learning has been handed down from generation to generation. The learning pattern remains sustainable in Sri Lanka even today.

The traditional textile industry has gradually developed into an industry that caters to the needs of the global context today. The end of the cultural revolution and the introduction of country-wide economic reforms, which saw the return of private enterprise, marked the second half of the 1970s as a period of change. In today's modern market, consumer demands and needs are constantly changing. New processes and steps are needed to keep up with these changes. The apparel industry of Sri Lanka has understood and long been practising that SDGs offer a common framework to increase the industry's own goals to address significant challenges in the global textile industry.

Through the analysis of the Textiles of Sri Lanka - Sustainable Innovations and Practices, this book explores the paths and methods of approaching Sustainable Development Goals. I would like to describe the diverse aspects to emphasise the uniqueness of the textiles of Sri Lanka and retain the interest and attention of many readers in this book.

The chapter "Sustainable Development Approaches for Sri Lankan Textile and Apparel Industry" overviews strategies for social responsibility and environmental impact reduction that the sector can implement. The chapter ""The Quality that Comes from Oral Tradition is the Trend and Luxury." Indigenous Hand-Loom Textiles Show Potential Towards Sustainable Development" discusses how oral tradition preserves the distinctive identity of indigenous cultural textiles, demonstrating how tradition-derived quality is both luxurious and in style while offering guidance on sustainable growth. This is the first study to investigate the so-called oral tradition as the caretaker of identity in indigenous cultural textiles, showing the potential for sustainable survival and sustainable development. The chapter "Heritage of Living Entity and Living Skills: Advancing Sustainability Goals 1 and 4 in Handloom Textiles" identifies how handloom textile weaving heritage prevails as a living entity and living skills in the Southern Province. Despite the challenges posed by the current economic situation and administrative shortcomings, the Department and its management recognise and value the weavers and the broader workforce for their significant economic contributions. It underscores the importance of collaborating with universities to leverage their educational resources and training for the weavers. This partnership also facilitates the dissemination of information to the broader local community. The chapter "The Potential of Utilizing Sri Lankan Banana Agro-Waste to Generate Sustainable Textiles" provides available data on banana varieties of Sri Lanka and emphasises potential opportunities for extracting fibre from common varieties, their physical, chemical, and morphological characteristics, and possible textile applications and related government interventions. This is the only research providing genetic features and identification details of Sri Lankan banana cultivars. Also, the research provides specific physio-chemical and morphological characteristics of Sri Lanka banana fibres. Moreover, the research provides sophisticated and sustainable opportunities for banana fibre utilisation in the textile industry. Further, the chapter is inimitable as it also provides the provisions and interventions taken by the government to facilitate entrepreneurial opportunities linked to the research. The chapter "Achieving Sustainable Development Goals: The Impact of the Apparel and Textile Industry on the Eradication of Poverty in Sri Lanka" investigates the impact of the apparel sector on the reduction of income poverty in Sri Lanka, explores the potential impact of the apparel and textile industry on eradicating multidimensional poverty in Sri Lanka, and examines the role of the apparel sector on fostering female empowerment in the country. The findings indicate that over one-third of the sample lives below the income poverty threshold, with nearly 20 per cent experiencing multidimensional poverty. The apparel and textile industry significantly contributes to reducing income and multidimensional poverty in the country. The research emphasises that as the leading sector that energises the Sri Lankan economy, the textile and apparel sector has silently contributed to eradicating rural poverty in Sri Lanka. The chapter "As Good as New: A Study on Sri Lanka Consumers' Secondhand Fashion Consumption" explores the concept of second-hand fashion consumption, which is not well-known in the Sri Lankan context. The research demonstrates five values that second-hand consumers consider: price consciousness, emotional bond consciousness, comfortability consciousness, quality and brand consciousness, and social and environmental consciousness. The research demonstrates non-second-hand fashion consumers' consideration values of social status consciousness, hygienic consciousness, comfortability consciousness, quality and brand consciousness, and self-expressive consciousness. The findings indicated that consumer purchase and word-of-mouth intentions grew as emotional bonds increased. This research is one of the first studies to investigate second-hand fashion consumption amongst Sri Lankan fashion consumers, consequently offering valuable insights into Sri Lankan fashion retailers' sustainability strategies. The chapter "The Use of ISO 14062:2002 Sustainable Design Strategies of the Designers to Address SDG 12: Ensure Sustainable Consumption and Production Patterns in the Export Apparel Industry in Sri Lanka" investigates the impact of the export apparel industry in Sri Lanka on sustainable development, ensuring sustainable consumption and production patterns to concise possible regulatory guidelines and suggestions for the Sri Lankan apparel industry. It reveals that interpreting sustainable design strategies from a profit-oriented perspective can attract the mass-market apparel industry to actual, sustainable initiatives to achieve SDG12 more realistically. This chapter presented the only research using ISO 14062:2022 sustainable design strategies of designers in the Sri Lankan export apparel industry to address SDG 12: Sustainable Consumption and Production Patterns. The chapter "Sustainability and Textile Testing" explains that testing plays a significant role in ascertaining the suitability of materials and processes regarding their sustainability and environmental impact. This chapter discusses various tests to ascertain the sustainability of textile materials and related processes. It also seeks to discuss the actual testing itself in terms of its sustainability. The chapter "An Investigation of Fashion for Career Development in Women's Role in the Apparel Industry in Sri Lanka" examines how fashion supports women's career development in the Sri Lankan apparel industry. The study focuses on women's career development and empowerment in the Sri Lankan apparel industry by thoroughly analysing the relationship between fashion preferences, professional image, and career advancement for women in Sri Lanka's garment industry. The chapter "Eco-Friendly Dyeing Techniques for Banana Fibers from *Banana Application in the Sri Lankan Context.* Promoting Sustainable Coloration Methods" emphasises eco-friendly dyeing techniques. The chapter promotes sustainable colouration methods by prioritising eco-friendly dyeing techniques. Also, the chapter presents practical applications and results. The chapter goes beyond theoretical concepts by providing practical insights and results. It explores the extraction methods of natural dyes, optimisation of dye bath conditions, and evaluates the properties of dyed banana fibres, such as lightfastness, wash fastness, and eco-toxicity. Readers can gain valuable knowledge and guidance for implementing these eco-friendly dyeing techniques. The chapter discusses the contribution of the sustainable fibre and textile industry. The chapter aligns with the industry's growing demand for sustainable and eco-friendly practices. By emphasising natural dye sources and minimising harmful chemicals, it offers a valuable resource for textile practitioners, designers, and researchers seeking to create environmentally conscious textiles. Overall, the chapter's unique selling points lie in its regional focus, emphasis on eco-friendly practices, practical application, and alignment with the sustainability trends in the fashion industry. These aspects make it a compelling and valuable contribution to the field. The chapter "Role of Traditional Crafts in Sustainable Development: Reflections on Dumbara Weaving" analyses the relationship between traditional crafts and sustainable development and identifies how those crafts can contribute to the long-term continuation of rural artisanal livelihood practices. The study proposes that Dumbara weaving and its endurance over time have been through the embeddedness of their livelihood within social institutions such as family and kinship and through the non-kin artisan-patron relationships, thereby contributing to achieving Sustainable Development Goals. It underlines the sixteenth of the UN's Sustainable Development Goals, which are to respect and promote cultural diversity, facilitate cultural understanding and peace, reduce inequalities, and promote the rights of marginalised groups.

Thank you for your expert knowledge, deep analysis, and wise judgement. Sri Lanka, as an Island nation, has identical approaches to achieving Sustainable Development Goals. Multi-cultural, religious, and diverse social interactions analyse unique patterns in approaching global Sustainable Development Goals.

Prof. Gayathri Madubhani Ranathunga
Department of Textile and Apparel Engineering
Fashion Design and Product Development Degree Course
Faculty of Engineering
University of Moratuwa
Moratuwa, Sri Lanka

Contents

Editor and Contributors

About the Editor

Gayathri Madubhani Ranathunga is a Professor (ON MERIT) in the Department of Textile and Apparel Engineering, Faculty of Engineering, University of Moratuwa, Sri Lanka. She obtained Bachelor of Arts (Hons) in Fine Arts from the Faculty of Humanities, University of Kelaniya, Sri Lanka, in 2003. She earned her post-graduate qualification of Doctor of Philosophy in Integrated Design at the University of Moratuwa in April 2012. She obtained a post-graduate certificate in learning and teaching art, design, and communication (cltad) from the University of Arts London, UK, in 2006. Prof. Gayathri joined the University of Moratuwa in 2004. She teaches histories and the fashion of costumes around the world and in Sri Lanka, as well as the art and tradition of Sri Lanka and fashion illustrations. Her research interests are Cultural studies in Dress and Fashion, socio-cultural, political, and religious Influences on costume and fashion, Material culture analysis in historical costumes, Intangible heritage, indigenous skills in textiles and sustainable survival and sustainable development, and fashion Illustrations. She has published many research papers, conference proceedings, presentations, abstracts, and book chapters in international and national research journals and books. Prof. Gayathri has authored two scholarly books, Fashions of Royalty: The Kandyan Kingdom of Sri Lanka and Art and Tradition of Sri Lanka Volume 01, Music of Sri Lanka. She is a co-author of the textbook "The Buddhist Robe": The Path Pointing to Natural Dyes and the Possible Establishment of a Natural Dyeing Industry in the Apparel Sector of Sri Lanka with special reference to the Buddhist Robe, published in 2019. She disseminates her knowledge and expertise in developing the country's secondary and tertiary education in the fashion design stream, with 20 years of expertise in the field. She is a committee member of Subject Benchmark for Design, appointed by the University of Grants Commission of Sri Lanka. She served as a member of the Institutional review panel organized by the Ministry of Higher Education and Research. She has provided her services as a consultant, editor, unit author, and external examiner to the Bachelor of Industrial Studies Fashion Design and Product

Development of the Open University of Sri Lanka, Nawala. Also, she was a member of the National Competency Standard for Fashion Designing Technology Development group at the University of Vocational Technology, Kandawala, Rathmalana, Sri Lanka. She is a consultant at the National Institute of Education, Maharagama, Sri Lanka, giving her expertise in the development of a syllabus of Fashion Design for the programme of '13 years of guaranteed Education'. She is a jury member of the National and Provincial Judge board of "*ShilpaAbhimani*" Presidential Awards in Handicrafts Competition, organized by the National Crafts Council, Ministry of Industry and Commerce. She is also a jury member of the National and Provincial Judge Board '*Ransalu*', National Textile Exhibition held by the Department of Textile Industries. She is also a consultant at the National Museum of Colombo, Sri Lanka.

Contributors

H. R. Achini Ranaweera Department of Textile and Apparel Engineering, University of Moratuwa, Moratuwa, Sri Lanka

Wasantha Athukorala Department of Economics and Statistics, University of Peradeniya, Peradeniya, Sri Lanka

Subashini Balakrishnan Department of Textile and Apparel Technology, Faculty of Engineering Technology, The Open University of Sri Lanka, Nawala, Nugegoda, Sri Lanka

K. P. N. Bhagya Department of Textile and Apparel Engineering, University of Moratuwa, Moratuwa, Sri Lanka

Sulari De Silva Department of Textile and Apparel Engineering, Faculty of Engineering, University of Moratuwa, Moratuwa, Sri Lanka

U. S. W. Gunasekera Department of Textile and Apparel Engineering, University of Moratuwa, Moratuwa, Sri Lanka

Rangika U. Halwatura Department of Civil Engineering, University of Moratuwa, Moratuwa, Sri Lanka

M. A. H. S. Hansadhi Faculty of Graduate Studies, University of the Visual and Performing Arts, Colombo, 07, Sri Lanka

U. I. Jayalath Department of Textile and Apparel Engineering, University of Moratuwa, Moratuwa, Sri Lanka

Muditha Karunarathna Department of Economics and Statistics, University of Peradeniya, Peradeniya, Sri Lanka

P. V. M. Karunaratne Department of Textile and Apparel Engineering, University of Moratuwa, Moratuwa, Sri Lanka

A. M. Lakshani Imeshika Department of Textile and Apparel Engineering, University of Moratuwa, Moratuwa, Sri Lanka

A. C. S. I. Mumthas Sri Lanka Institute of Textile and Apparel, Ratmalana, Sri Lanka

Prabod Dharshana Munasinghe Department of Textile and Apparel Engineering, University of Moratuwa, Moratuwa, Sri Lanka

S. N. Niles Department of Textile and Apparel Engineering, University of Moratuwa, Moratuwa, Sri Lanka

G. M. Ranathunga Department of Textile and Apparel Engineering, University of Moratuwa, Moratuwa, Sri Lanka

K. G. R. Samarawickrama Department of Textile and Apparel Engineering, University of Moratuwa, Moratuwa, Sri Lanka

U. G. Samudrika Wijayapala Department of Textile and Apparel Engineering, University of Moratuwa, Moratuwa, Sri Lanka

P. H. Thamoda Geegamage Department of Textile and Apparel Engineering, University of Moratuwa, Moratuwa, Sri Lanka

C. Thenuwara Department of History and Theory of Arts, Faculty of Visual Arts, University of the Visual and Performing Arts, Colombo, 07, Sri Lanka

Thushari N. Vidanage Department of Economics and Statistics, University of Peradeniya, Peradeniya, Sri Lanka

Lakshika C. Weragoda Department of Economics and Statistics, University of Peradeniya, Peradeniya, Sri Lanka

G. L. D. Wickramasinghe Department of Textile and Apparel Engineering, University of Moratuwa, Moratuwa, Sri Lanka

D. S. Wijerathne Department of Textile and Apparel Technology, The Open University of Sri Lanka, Nawala, Nugegoda, Sri Lanka

Sustainable Development Approaches for Sri Lankan Textile and Apparel Industry

U. G. Samudrika Wijayapala and K. G. R. Samarawickrama

Abstract The Sri Lankan textile and apparel industry is a significant contributor to the country's economy, but it also faces challenges related to sustainable development. This chapter explores sustainable development approaches that the industry can adopt to reduce environmental impact and promote social responsibility. The study begins by examining the current state of the textile and apparel industry in Sri Lanka and identifying its sustainability challenges, including water scarcity, carbon emissions, and labour practices. The chapter then discusses sustainable development approaches such as circular economy, eco-design, and green supply chain management, which can help address these challenges. The circular economy model emphasizes resource efficiency and waste reduction, eco-design incorporates sustainability into the design process, and green supply chain management promotes sustainability throughout the entire supply chain. These approaches can enable the textile and apparel industry in Sri Lanka to become more environmentally and socially sustainable, while also improving economic performance. Overall, the adoption of sustainable development approaches is critical for the long-term success of the Sri Lankan textile and apparel industry.

Keywords Sustainable development · Environmental impact · Circular economy · Eco-design · Green supply chain management

1 Introduction

The Sri Lankan textile and apparel industry has long been a cornerstone of the country's economic growth, serving as a critical driver of exports and employment. However, this industry is undergoing a profound transformation in the face of global

U. G. Samudrika Wijayapala (✉) · K. G. R. Samarawickrama
Department of Textile and Apparel Engineering, University of Moratuwa, Moratuwa, Sri Lanka
e-mail: samu@uom.lk

K. G. R. Samarawickrama
e-mail: samrawickramakgr.20@uom.lk

G. M. Ranathunga (ed.), *Textiles of Sri Lanka*, SDGs and Textiles,
https://doi.org/10.1007/978-981-97-0569-6_1

sustainability challenges. Sustainable development approaches are emerging as a critical imperative, reshaping the way Sri Lanka's textile and apparel sector operates and thrives in the twenty-first century. Sustainability in this context transcends mere profitability and production. It encompasses a holistic commitment to environmental conservation, ethical labour practices, and social responsibility, all while ensuring the industry's long-term economic viability [59]. The Sri Lankan textile and apparel industry is at a pivotal crossroads, where sustainable practices are no longer optional but essential for its survival and global competitiveness. The Sri Lankan textile and apparel industry plays a pivotal role in the country's economy, contributing significantly to exports and providing employment opportunities for a large workforce. However, in an era marked by growing environmental concerns, ethical labour practices and evolving consumer preferences, there is an urgent need for the industry to adopt sustainable development approaches [2].

Sustainability in this context encompasses environmental stewardship, social responsibility, and economic viability. The sustainable development approaches in the Sri Lankan textile and apparel industry are not merely a choice but a necessity. It requires a fundamental shift in the way the industry operates, from raw material sourcing to production processes and supply chain management. This shift must prioritize long-term sustainability over short-term gains, fostering innovation and aligning with international standards and best practices [3, 5]. Sri Lanka has the opportunity to position itself as a global leader in sustainable textile and apparel manufacturing. By adopting and promoting sustainable practices, the industry can enhance its competitiveness, gain access to environmentally conscious markets, and contribute to a more responsible and equitable global supply chain. This journey towards sustainability is multifaceted, encompassing efforts to reduce environmental impact, uphold labour rights, and engage in ethical business practices. It also involves educating and training the workforce, fostering collaboration between industry stakeholders and integrating sustainability principles into educational curricula. In this series of articles, we will explore the various dimensions of sustainable development for the Sri Lankan textile and apparel industry [6, 17]. We will delve into strategies to reduce carbon emissions, minimize water usage, and manage waste responsibly. We will also examine the importance of fair labour practices, supply chain transparency, and ethical sourcing. Furthermore, we will explore how educational curriculum enhancement can play a pivotal role in shaping a sustainable future for the industry. The Sri Lankan textile and apparel sector sees sustainability as a strategic objective rather than a trendy concept. Through thoughtful and concerted efforts, the industry can secure its future and contribute positively to the well-being of the planet and its people [7]. In this chapter, we discuss the sustainability challenges and sustainable development approaches for sustainable development approaches for Sri Lankan textile and apparel industry.

2 Sustainability Challenges

The Sri Lankan textile and apparel industry has long been a cornerstone of the country's economic growth, serving as a key driver of exports and employment. However, this industry is undergoing a profound transformation in the face of global sustainability challenges. Sustainable development approaches are emerging as a critical imperative, reshaping the way Sri Lanka's textile and apparel sector operates and thrives in the twenty-first century. Sustainability in this context transcends mere profitability and production. It encompasses a holistic commitment to environmental conservation, ethical labour practices, and social responsibility, all while ensuring the industry's long-term economic viability. The Sri Lankan textile and apparel industry is at a pivotal crossroads, where sustainable practices are no longer optional but essential for its survival and global competitiveness. Sustainability challenges are complex and pressing issues related to the environment, society, and the economy that hinder the ability of individuals, communities, organizations, and nations to achieve long-term well-being while also ensuring that resources are preserved for future generations [8, 10]. These challenges are typically characterized by their multi-faceted nature and the need for comprehensive, integrated solutions. Indeed, the Sri Lankan textile and apparel industry faces some sustainability challenges as it seeks to adopt sustainable development approaches. Water scarcity is a growing sustainability challenge impacting industries worldwide, including Sri Lankan textile and apparel. This sector relies heavily on water throughout its supply chain, from the cultivation of raw materials to the manufacturing processes. As water resources become scarcer due to climate change and increasing demand, addressing this challenge is paramount for the industry's sustainability. Water scarcity arises from the imbalance between water supply and demand, and it is exacerbated by factors such as population growth, industrialization, and changing weather patterns. Like many other regions, Sri Lanka faces the risk of dwindling freshwater resources. For the textile and apparel industry, which relies on substantial water usage for dyeing, finishing, and other processes, water scarcity substantially threatens its operations, the environment, and local communities [9, 11].

Carbon emissions are a significant sustainability challenge facing industries worldwide, and the Sri Lankan textile and apparel industry is no exception. This sector primarily contributes to greenhouse gas emissions through energy-intensive processes and transportation. As the global community grapples with the urgent need to address climate change, mitigating carbon emissions has become a crucial aspect of sustainability. Carbon emissions, primarily in the form of carbon dioxide (CO_2) and other greenhouse gases, are a key driver of climate change. Sri Lanka, along with the rest of the world, is experiencing the adverse impacts of rising temperatures, extreme weather events and shifting climate patterns. The textile and apparel industry is a notable contributor to carbon emissions due to its reliance on energy-intensive processes, such as dyeing and finishing, and the extensive transportation of goods across the supply chain [12, 13].

Chemical management and pollution pose significant sustainability challenges, particularly in industries like the Sri Lankan textile and apparel sector, which rely on various chemicals in their manufacturing processes. The improper use, disposal, and management of chemicals can lead to environmental contamination, harm human health, and undermine the industry's sustainability efforts. Chemicals are integral to textile and apparel manufacturing, from dyes and finishes to treatments for fabric and garments. However, the improper handling, release, and disposal of these chemicals can have severe environmental and health consequences. Pollution, such as water and soil contamination, directly results from chemical mismanagement. Additionally, worker safety and well-being can be compromised if proper precautions are not in place. These challenges raise ethical and environmental concerns that must be addressed to achieve sustainability [14, 15].

Waste management is a critical sustainability challenge faced by industries worldwide, and the Sri Lankan textile and apparel industry is no exception. This sector generates a significant amount of waste, including off-cuts, defective products, and discarded garments. Inefficient waste management can lead to environmental pollution, resource depletion, and reputational damage. Addressing waste management challenges is essential for the industry to align with sustainable practices. The issue of waste management can be tackled to promote sustainability in the textile and apparel industry. The textile and apparel industry produces substantial waste during various stages of production, from raw material processing to finished product manufacturing. Inadequate waste management practices can result in environmental harm, such as landfills overflowing with textile waste and pollution from the disposal of hazardous materials. Sustainable waste management minimizes these adverse effects [20, 24].

Supply chain transparency is fundamental to sustainability, particularly in industries like the Sri Lankan textile and apparel sector. This industry's global supply chain involves various stages, from sourcing raw materials to manufacturing and distribution. Ensuring transparency throughout this complex web of suppliers and subcontractors is essential for addressing sustainability challenges. The textile and apparel supply chain is often intricate, involving multiple tiers of suppliers, subcontractors, and factories across different countries. Ensuring transparency at each stage is challenging but vital for addressing sustainability concerns. Issues such as unethical labour practices, environmental degradation and harmful chemicals can arise when supply chains lack transparency [21, 22].

Skills and education play a crucial role in promoting sustainability within the Sri Lankan textile and apparel industry. Like many others, this sector faces significant sustainability challenges from environmental impact to labour practices. The textile and apparel industry requires diverse skills, from design and production to supply chain management and sustainability practices. The challenge lies in ensuring that the workforce possesses the knowledge and competencies needed to drive sustainability initiatives effectively. Education and training must keep pace with evolving sustainability standards and best practices [25, 28].

Economic sustainability is critical to overall sustainability in any industry, including the Sri Lankan textile and apparel industry. It encompasses the ability to balance economic viability with social and environmental responsibility. Economic sustainability challenges arise when businesses must navigate the complexities of reducing their environmental impact and promoting ethical labour practices while maintaining profitability. The textile and apparel industry faces a unique challenge in balancing sustainability with economic sustainability. While transitioning to sustainable practices often requires upfront investments in eco-friendly technologies, responsible sourcing, and ethical labour practices, the immediate costs can strain profit margins. Economic sustainability challenges emerge as businesses must find ways to absorb these initial costs while ensuring that long-term economic benefits outweigh them [29, 30]

The adoption of a circular economy is a transformative concept that challenges traditional linear models of production and consumption. In industries like the Sri Lankan textile and apparel sector, circular economy adoption presents both opportunities and challenges for sustainability. The shift towards circularity involves rethinking product design, manufacturing processes, and waste management. Circular economy adoption challenges traditional business models that prioritize linear production, consumption, and disposal. In the textile and apparel industry, where fast fashion has been prevalent, transitioning to a circular model means extending product lifecycles, reusing materials, and minimizing waste. This shift requires significant changes in product design, manufacturing, supply chains, and consumer behaviour [36, 38].

3 Sustainable Development Approaches

The Sri Lankan textile and apparel industry stands at a pivotal juncture where adopting sustainable development approaches is not just a choice but a necessity. This industry, a significant contributor to the country's economy, has experienced substantial growth over the years. However, this growth has come at a cost, with environmental concerns and ethical labour practices increasingly pressing. Sustainable development approaches represent a comprehensive strategy aimed at addressing these challenges while ensuring the industry's long-term viability. Sustainability in this context encompasses a broad spectrum of objectives, including reducing environmental impact, promoting ethical labour practices, optimizing resource usage, and fostering social responsibility. It calls for a fundamental shift in the way the industry operates, from the sourcing of raw materials and production processes to consumer behaviour and waste management. These approaches are essential for safeguarding the environment and improving the well-being of the workforce but also for maintaining the industry's competitiveness in a global market increasingly focused on sustainability. In exploring sustainable development approaches for the Sri Lankan textile and apparel industry, we will explore various strategies and initiatives to achieve a harmonious balance between economic growth, environmental responsibility, and ethical practices. These approaches present challenges and offer significant opportunities for innovation, efficiency, and long-term prosperity. Ultimately, they are a blueprint for transforming an industry traditionally associated with fast fashion and high resource consumption into a beacon of sustainable and responsible business practices [19, 55].

Environmental sustainability is a cornerstone for improving sustainable development approaches in the Sri Lankan textile and apparel industry. This sector, while contributing significantly to the country's economy, has faced mounting challenges related to its environmental impact. Embracing environmental sustainability addresses these concerns and positions the industry for long-term resilience and competitiveness. One of the key aspects of environmental sustainability is resource efficiency. The textile and apparel industry is known for its high water and energy consumption. To address this, companies can invest in technologies that optimize resource usage. Water recycling systems, energy-efficient machinery, and reduced material waste are essential steps. These practices minimize environmental impact and reduce operational costs [39]. Sustainable sourcing is a critical strategy for enhancing sustainable development approaches in the Sri Lankan textile and apparel industry. Sustainable sourcing offers a pathway to achieving both environmental and social responsibility while ensuring the industry's long-term viability. Sustainable sourcing involves procuring raw materials that have a minimal environmental impact. This includes prioritizing organic cotton, recycled fibres and eco-friendly dyes. The industry can significantly reduce its carbon footprint and minimize its effects on ecosystems by choosing materials produced using sustainable agricultural practices and eco-friendly manufacturing processes. Sourcing from local suppliers and those who adhere to ethical labour practices is a crucial aspect of sustainability. By collaborating with suppliers who provide fair wages, safe working conditions, and adhere to ethical labour standards, the industry can ensure that its operations promote social responsibility and worker well-being [41].

Waste reduction is a crucial component of sustainable development approaches in the Sri Lankan textile and apparel industry. This industry is notorious for generating significant amounts of waste, including fabric off-cuts, unsold inventory, and discarded garments. Implementing effective waste reduction strategies minimizes the environmental impact and contributes to cost savings and overall sustainability. One of the primary sources of waste in the textile and apparel industry is inefficient material usage. Through meticulous pattern layouts, cutting techniques, and fabric optimization, companies can reduce the amount of textile waste generated during production. Overproduction is common in fast fashion, leading to excessive waste [42]. Adopting a more demand-driven approach, where production is aligned with actual customer orders, can significantly reduce overproduction. Another effective waste reduction strategy is repurposing and recycling textile scraps and unsold inventory. Rather than discarding these materials, they can be creatively repurposed into new products or recycled into new textiles. This reduces waste and extends the lifecycle of materials and aligns with circular economy principles. Donating unsold inventory to charitable organizations or implementing take-back programs that allow customers to return used clothing for resale or recycling can minimize waste and promote social responsibility. These initiatives not only reduce landfill waste but also contribute to community well-being. Educating consumers about the importance of responsible consumption and disposal of textiles and apparel is crucial. By raising awareness about the consequences of fast fashion and encouraging consumers to recycle or donate clothing, the industry can foster a culture of sustainability. Embracing circular economy principles, such as designing products for durability, repairability, and recyclability, is pivotal. This approach extends the life of products and materials, reducing the need for new resources and minimizing waste generation. Ensuring compliance with waste disposal regulations is essential. Proper waste management practices, such as recycling and safe disposal, prevent environmental contamination and legal consequences [4, 44]

The adoption of green technologies is a pivotal strategy for improving sustainable development approaches in the Sri Lankan textile and apparel industry. Historically associated with resource-intensive and environmentally harmful practices, this sector is undergoing a transformation driven by the integration of eco-friendly technologies. Embracing these innovations reduces the industry's environmental footprint and enhances its long-term sustainability and competitiveness. Green technologies in the form of energy-efficient machinery and equipment are revolutionizing textile and apparel production. Companies can significantly reduce energy consumption by replacing older, energy-intensive machinery with more efficient models [45, 46]. This lowers operational costs and mitigates greenhouse gas emissions, contributing to environmental sustainability. Harnessing renewable energy sources like solar and wind power is another facet of green technology adoption. By installing solar panels or wind turbines, textile and apparel factories can generate their own clean energy, reducing their reliance on fossil fuels. This reduces carbon emissions and insulates businesses from energy price fluctuations. Traditional dyeing and printing processes in the industry often involve the use of chemicals and large volumes of water. Green technologies offer alternatives such as digital printing and waterless dyeing processes. These methods drastically reduce water consumption, chemical usage, and wastewater discharge, aligning with environmental sustainability goals. Innovations in sustainable materials, such as recycled or bio-based fabrics, are becoming increasingly prevalent. These materials reduce the environmental impact of textile production and cater to consumers growing demand for eco-friendly products. Advanced water recycling systems are emerging as essential green technologies in the textile and apparel industry. These systems capture and treat wastewater, allowing it to be reused in production processes. Green technologies also include sophisticated monitoring and control systems that optimize resource usage. Sensors and automation technology can regulate machine performance, reduce energy consumption during non-production hours, and minimize material waste, enhancing overall resource efficiency. Sustainable packaging solutions, such as biodegradable or recyclable materials, are integral to green technology adoption. By using eco-friendly packaging, companies reduce their contribution to plastic pollution and promote responsible disposal practices [31, 47].

Regulatory compliance plays a crucial role in improving sustainable development approaches in the Sri Lankan textile and apparel industry. As the industry faces mounting environmental and ethical challenges, adherence to local and international regulations is a legal requirement and a fundamental step towards promoting responsible practices and enhancing long-term sustainability. Compliance with these regulations requires textile and apparel companies to adopt eco-friendly practices, reduce pollution, and manage waste responsibly. For instance, adherence to limits on wastewater discharge and the proper disposal of hazardous chemicals are essential aspects of environmental compliance. By meeting these standards, companies minimize their environmental footprint and contribute to the preservation of local ecosystems. Beyond environmental concerns, the Sri Lankan textile and apparel industry must address labour-related issues. Compliance with ethical labour standards, such as fair wages, safe working conditions, and protection of workers' rights, is vital. International organizations like the International Labour Organization (ILO) set guidelines for fair labour practices. Ensuring workers are treated ethically aligns with global human rights principles and enhances the industry's social sustainability. Many textile and apparel companies seek certifications and labels that demonstrate their commitment to sustainability and regulatory compliance [43, 51].

These certifications, such as Fair Trade or Global Organic Textile Standard (GOTS), provide consumers with assurance and open access to eco-conscious markets. Compliance with the standards associated with these certifications necessitates adherence to strict environmental and ethical criteria. Transparency in reporting environmental and social performance is increasingly important. Many companies voluntarily disclose their sustainability efforts through annual reports or dedicated sustainability websites. This practice builds consumer trust and highlights areas where the industry meets or exceeds regulatory requirements. Regulatory compliance extends throughout the supply chain. Companies must ensure that their suppliers and subcontractors adhere to environmental and labour regulations. Regular audits and monitoring of supply chain partners are essential to maintain compliance and minimize risks associated with non-compliance. Collaborating with governmental bodies is crucial for promoting sustainable development. Textile and apparel industry associations can work closely with government agencies to develop supportive policies, incentives, and regulations encouraging sustainable practices. Government support can further drive positive change within the industry [51, 54]

Consumer awareness plays a pivotal role in improving sustainable development approaches in the Sri Lankan textile and apparel industry. As consumers become increasingly conscious of environmental and ethical issues associated with the fashion industry, businesses are compelled to respond by adopting sustainable practices. Consumer awareness campaigns are essential for educating the public about the environmental and social implications of their clothing choices. These campaigns can highlight the environmental impact of fast fashion, including water and resource consumption, pollution, and textile waste. By raising awareness about these issues, consumers are more likely to make informed, sustainable choices when purchasing clothing. As consumers become more informed about sustainability issues, there is a growing demand for eco-friendly and ethically produced textiles and apparel. This shift in consumer preferences incentivizes businesses to prioritize sustainable sourcing, production processes, and materials. Companies that respond to this demand can gain a competitive edge and contribute to industry-wide sustainability. Informed consumers demand business transparency. They want to know where and how their clothing is produced and whether ethical labour practices are upheld. This demand for transparency compels companies to improve supply chain visibility, traceability, and reporting. It encourages businesses to adhere to ethical labour standards and environmentally responsible practices to maintain consumer trust [53, 56].

Consumer awareness campaigns can lead to behavioural change, encouraging consumers to rethink their consumption habits. Concepts like "buy less, choose well, make it last" promote a more sustainable approach to fashion, emphasizing quality over quantity and reducing the frequency of disposal. Such shifts in consumer behaviour align with sustainability goals and reduce the industry's environmental impact. Informed consumers actively seek out and support sustainable brands [57]. Companies committed to responsible practices, ethical labour, and eco-friendly materials can attract loyal customers. This support drives market competition, incentivizing more businesses in the textile and apparel industry to adopt sustainable practices. Consumer awareness often translates into pressure and advocacy. In response to consumer demands, some consumers and organizations actively campaign for change within the industry. This advocacy can lead to policy changes, industry-wide initiatives, and increased scrutiny of unsustainable practices. Informed consumers are more likely to participate in sustainable initiatives such as clothing recycling programs, clothing swaps, and second-hand shopping. These activities promote circularity, reduce textile waste, and extend the lifespan of clothing items [23, 61].

Ethical labour practices are fundamental to improving sustainable development approaches in the Sri Lankan textile and apparel industry. Embracing ethical labour practices addresses these issues and contributes to the industry's long-term sustainability and reputation. Ethical labour practices demand safe and healthy work environments for textile and apparel workers. This includes proper ventilation, lighting, sanitation facilities, and measures to prevent accidents and injuries. By prioritizing worker well-being, businesses fulfil their moral obligations and reduce the risk of workplace-related incidents and associated costs. Ethical labour practices uphold workers' rights, including the right to fair wages, reasonable working hours, and freedom from discrimination and harassment. Ensuring fair compensation and adherence to legal working hours helps improve the livelihoods of workers and promotes social sustainability. Fostering a diverse and inclusive workplace enhances the industry's reputation and competitiveness. Ethical labour practices unequivocally reject child labour and forced labour. By conducting thorough supply chain assessments and implementing stringent monitoring mechanisms, the industry can identify and eliminate any instances of child labour or coercion in production. This not only aligns with global labour standards but also safeguards the well-being of vulnerable populations [60, 62]

Many textile and apparel companies seek Fair Trade and ethical certifications to demonstrate their commitment to responsible labour practices. These certifications assure consumers of ethical treatment and promote responsible and transparent supply chains. This helps businesses build trust and loyalty among consumers. Collaboration with Labor Organizations: Collaborating with labour organizations and unions can be instrumental in improving labour conditions. Businesses can address concerns and find mutually beneficial solutions by engaging in open dialogue and negotiation with workers representatives. This collaboration fosters positive labour relations and contributes to workplace stability. Ethical labour practices involve investing in worker training and development programs. These initiatives enhance workers' skills, employability, and job satisfaction. Companies benefit from a more skilled workforce, reduced turnover, and improved productivity. Ethical labour practices extend throughout the supply chain. Companies must conduct regular supplier audits to ensure subcontractors and suppliers adhere to ethical labour standards. Holding suppliers accountable for ethical practices reinforces responsible sourcing and production [65, 67].

Consumer education is a potent tool for improving sustainable development approaches in the Sri Lankan textile and apparel industry. As consumers become increasingly aware of environmental and ethical concerns associated with the fashion industry, educating them about responsible choices can drive positive change in the industry. Consumer education initiatives can highlight the significant environmental impact of the textile and apparel industry. This includes the industry's contribution to water pollution, carbon emissions, and textile waste. By raising awareness of these issues, consumers are more likely to make eco-conscious choices when purchasing clothing [16].

Promoting ethical consumption is a powerful strategy for improving sustainable development approaches in the Sri Lankan textile and apparel industry. This approach focuses on raising awareness and encouraging consumers to make choices that align with ethical principles, such as fair labour practices and social responsibility. One of the fundamental aspects of promoting ethical consumption is consumer education. Initiatives and campaigns can inform consumers about the ethical challenges in the textile and apparel industry, including issues related to worker conditions, wages, and human rights. Educating consumers about these concerns makes them more likely to make informed choices when purchasing clothing. Ethical consumption encourages consumers to support brands and businesses that prioritize fair labour practices and uphold workers' rights. Consumers are encouraged to research and choose products from companies committed to ethical labour standards. This support sends a clear message to the industry that ethical practices are valued and rewarded. Ethical consumption promotes transparency within the industry. Consumers are encouraged to seek out brands that provide information about their supply chains, sourcing practices, and ethical commitments. This transparency allows consumers to make informed decisions about the brands they support and encourages businesses to be accountable for their practices [68, 69].

Consumers engaged in ethical consumption may boycott brands or products associated with unethical practices. This can pressure companies to improve their ethical standards to avoid losing market share. Fair trade and ethical certifications serve as indicators of ethical consumption. These certifications ensure that workers are treated fairly, receive fair wages, and work in safe conditions. Encouraging consumers to look for and support products with these certifications helps promote ethical practices within the industry. Ethical consumption often aligns with sustainable fashion movements such as the Fashion Revolution. These movements call for transparency, accountability, and ethical practices in the fashion industry. Engaging in such movements amplifies the message of responsible fashion and pressures businesses to improve their ethical standards. Ethical consumption fosters a culture of responsibility among consumers. It encourages them to consider their clothing choices' broader social and environmental impacts. This shift in mindset can lead to long-term changes in consumer behaviour and industry practices [52, 66].

Encouraging responsible consumption habits is vital for improving sustainable development approaches in the Sri Lankan textile and apparel industry. This approach focuses on promoting mindful and sustainable consumer behaviour that reduces waste, supports ethical practices, and contributes to the industry's long-term sustainability. Responsible consumption habits emphasize choosing quality over quantity. Consumers are encouraged to invest in well-made, durable clothing that withstands wear and tear, reducing the need for frequent replacements. This approach not only reduces textile waste but also aligns with the principles of sustainability. Responsible consumers are urged to extend the lifespan of their clothing items. Proper care, maintenance, and repair of garments help ensure that they remain in use for extended periods. This reduces the disposal of clothing and minimizes the demand for new production. Responsible consumption discourages impulse purchases driven by fast fashion trends [18, 37].

Consumers are encouraged to think carefully about their clothing needs and make deliberate, thoughtful choices. This habit reduces overconsumption and the subsequent disposal of lightly worn clothing. Responsible consumers actively seek out and support sustainable and ethical brands. They prioritize products made from eco-friendly materials, embrace ethical production practices, and support companies that align with their values. This support motivates businesses to adopt sustainable practices to meet consumer demand. Responsible consumption habits include participating in second-hand shopping and clothing swaps. These activities promote circular fashion by giving pre-owned clothing a second life. Consumers can reduce textile waste and enjoy affordable, sustainable fashion options. Responsible consumers consider the responsible disposal of clothing they no longer need. This may involve donating to charitable organizations, recycling textiles, or selling clothing through second-hand platforms. It ensures that clothing items are diverted from landfills and can benefit others. Consumer education programs play a crucial role in promoting responsible consumption habits. They inform consumers about the environmental and social impacts of the fashion industry, encouraging them to make more sustainable choices. Responsible consumption habits are about changing consumer mindsets. Consumers are encouraged to view clothing as long-term investments rather than disposable commodities. This shift in perspective has the potential to reshape the entire fashion industry by reducing its emphasis on fast fashion [48, 63].

Educating consumers about sustainable materials and production processes empowers them to make informed choices. Consumers can seek products made from eco-friendly materials, such as organic cotton or recycled fibres, and support brands that employ sustainable production methods like low-impact dyeing. Consumer education promotes transparency within the industry. Consumers are encouraged to seek out brands that provide information about their supply chains, sourcing practices, and environmental efforts. Educated consumers are more likely to participate in second-hand shopping, clothing swaps, and clothing rental services. These activities promote circular fashion, reduce textile waste, and extend the lifecycle of clothing items. Informed consumers often become advocates for sustainable fashion. They may support and promote brands that align with their values while pressuring others to improve their sustainability practices. Consumer pressure can lead to policy changes and industry-wide initiatives. Consumer education encourages participation in sustainability movements such as the Fashion Revolution, which calls for transparency and accountability in the fashion industry. Engaged consumers amplify the message of responsible fashion and hold businesses accountable [34, 50].

Innovation in sustainable materials is a pivotal strategy for improving sustainable development approaches in the Sri Lankan textile and apparel industry. This industry, which traditionally relies on resource-intensive and environmentally harmful materials, can significantly reduce its environmental footprint and enhance sustainability by embracing innovative materials and production processes. One key innovation is the adoption of organic and recycled fibres. Organic cotton, for instance, is grown without synthetic pesticides or genetically modified seeds, reducing chemical use and environmental impact. Similarly, recycled fibres are derived from post-consumer textile waste, diverting materials from landfills and reducing the need for virgin resources. Innovations in bio-based materials, such as fabrics derived from agricultural residues or fungi, are gaining traction. These materials are renewable and often have a lower environmental footprint compared to traditional textiles. Incorporating these alternatives into production can significantly reduce resource consumption and pollution. Traditional dyeing and finishing processes are resource-intensive and can result in water pollution. Innovations in eco-friendly dyes and finishes use less water and fewer chemicals, reducing the environmental impact of textile production [32, 58]. These technologies conserve resources and improve the industry's environmental reputation. Technological innovation has led to smart fabrics that offer sustainability benefits. These fabrics can respond to environmental conditions, such as temperature or moisture, reducing the need for additional layers or energy-consuming heating and cooling systems. They enhance comfort while promoting resource efficiency. Some innovative materials are designed to be biodegradable, breaking down naturally at the end of their lifecycle. This reduces textile waste and addresses the issue of garments ending up in landfills. Biodegradable textiles are particularly relevant in a circular fashion framework. Leather production has significant environmental and ethical challenges. Innovations in sustainable leather alternatives, such as lab-grown or plant-based materials, offer ethical and environmentally friendly options for the fashion industry. Some innovations involve turning waste materials into textiles, such as agricultural byproducts or food waste. These closed-loop systems reduce waste while creating new sustainable materials for the industry. Collaboration between industry players, research institutions, and government agencies is essential for advancing sustainable materials. Joint research projects can lead to breakthroughs in sustainable textile innovation and the development of eco-friendly materials [18, 58].

Leveraging advanced production technologies is crucial for enhancing sustainable development approaches in the Sri Lankan textile and apparel industry. This sector faces considerable environmental and ethical challenges, and innovation in production processes can lead to significant improvements in sustainability. Traditional dyeing methods in the textile industry consume vast amounts of water and chemicals, contributing to pollution and resource depletion. Advanced production technologies offer low-impact dyeing techniques significantly reducing water usage and chemical waste. Digital printing and waterless dyeing systems have emerged as environmentally friendly alternatives. Digital technologies enable designers to create and prototype clothing in virtual environments, reducing the need for physical samples and minimizing material waste. This accelerates the design process and lowers the industry's carbon footprint by reducing fabric and resource consumption. Automation and robotics have the potential to improve efficiency and precision in textile and apparel manufacturing. Automated sewing machines, for example, can reduce human labour while increasing the accuracy and consistency of stitching. This produces higher-quality products and reduces the industry's dependence on low-wage work, promoting ethical labour practices [57, 62].

The Internet of Things (IoT) and data analytics enable real-time monitoring of manufacturing processes. This data-driven approach helps identify inefficiencies and areas for improvement, leading to more sustainable manufacturing practices. For instance, energy and resource consumption can be optimized to reduce environmental impact. 3D printing technologies are increasingly being applied to textile production. These technologies allow for precise material allocation, minimizing waste. Customization and on-demand production also become more feasible, reducing overproduction and textile waste. Advanced production technologies like pattern nesting software can incorporate waste reduction techniques that optimize fabric utilization. This minimizes the amount of fabric discarded as scrap during cutting and contributes to resource efficiency. The technology helps identify areas where sustainability can be enhanced within the supply chain. Collaborations between textile and apparel companies and technology firms can drive innovation. Tech companies often bring expertise in data analytics, automation, and supply chain management that can be harnessed to develop sustainable production processes [10, 17].

Implementing sustainability in supply chains is a pivotal strategy for improving sustainable development approaches in the Sri Lankan textile and apparel industry. The industry's complex global supply chains encompass various stages, from raw material sourcing to final product distribution, and addressing sustainability at each step can have far-reaching positive impacts. Ethical Sourcing of Raw Materials: Sustainable supply chains begin with ethically sourcing raw materials. In Sri Lanka, for instance, cotton, one of the primary raw materials, can be sourced in ways that prioritize fair trade and responsible farming practices. By ensuring that raw materials are produced ethically, the industry contributes to improved labour conditions and reduced environmental harm. Transparency is a cornerstone of sustainable supply chains. Companies can map their supply chains to understand their products journey from inception to delivery. This transparency helps identify potential bottlenecks, inefficiencies, and areas for improvement in terms of sustainability. Regular audits and certifications of supply chain partners are crucial for ensuring adherence to ethical and environmental standards. These audits help identify non-compliance and encourage suppliers to rectify labour practices, waste management, and resource consumption issues [12, 32].

Collaboration between industry players and suppliers is essential for promoting sustainability. Joint initiatives, such as capacity-building programs and knowledge sharing, can help suppliers adopt sustainable practices, from ethical labour standards to resource-efficient production methods. Embracing circularity in supply chains can drastically reduce waste and resource consumption. Sri Lankan textile and apparel companies can explore options for recycling and reusing materials and products within their supply chains, reducing the need for virgin resources and minimizing waste disposal. Sustainable supply chains consider the environmental impact of logistics and transportation. Companies can adopt eco-friendly transportation options, optimize routes to reduce emissions, and explore alternative modes of transportation to lower their carbon footprint. Industry demand for sustainable materials can drive suppliers to innovate and provide eco-friendly alternatives. By prioritizing sustainable materials in their supply chain, businesses encourage suppliers to transition to more responsible practices. Consumers are increasingly demanding sustainability in the products they purchase. Aligning supply chains with consumer expectations and market trends can make businesses more competitive and resilient. Sustainable supply chains help meet this growing demand for ethical and environmentally friendly products [26, 64].

Leveraging consumer engagement platforms is vital for improving sustainable development approaches in the Sri Lankan textile and apparel industry. These platforms offer a direct channel for businesses to interact with consumers, educate them about sustainability, and encourage responsible choices. Consumer engagement platforms serve as educational hubs where companies can inform consumers about the textile and apparel industry's environmental and ethical challenges. Consumers better understand issues such as water scarcity, carbon emissions, ethical labour practices, and textile waste through informative content, including articles, videos, and infographics. Consumer engagement platforms enable businesses to promote sustainable choices. Companies can showcase eco-friendly and ethically produced clothing lines, highlighting organic or recycled materials and responsible production processes. This encourages consumers to make informed decisions when purchasing clothing. These platforms provide advice on responsible consumption habits. Tips on buying fewer but higher-quality garments, clothing care practices, and clothing repair and recycling options can all be featured. Such guidance empowers consumers to extend the lifespan of their clothing and reduce waste [26, 66]. Consumer engagement platforms can emphasize supply chain transparency and brand accountability. Companies can share information about their supply chain practices, materials sourcing, and efforts to reduce their environmental impact. This transparency builds trust with consumers, who increasingly seek brands that align with their values. Two-way communication on these platforms allows consumers to provide feedback and voice their concerns. Businesses can use this feedback to improve their sustainability practices, address consumer inquiries, and demonstrate their commitment to responsible practices. Consumer engagement platforms can promote circular fashion initiatives such as clothing rental, second-hand shopping, and clothing swaps. These activities reduce textile waste and encourage a more circular economy within the industry. These platforms can foster a sense of community among like-minded consumers who prioritize sustainability. Forums, discussion boards, or social media groups can facilitate discussions, knowledge sharing, and collective efforts to drive sustainability in the industry. Businesses can use these platforms to launch sustainability-focused campaigns and advocate for industry-wide changes. Engaging consumers in advocacy efforts, such as supporting fair labour practices or urging more sustainable supply chains, amplifies the industry's sustainability message [35, 69].

Government and industry collaboration is crucial to improving sustainable development in the Sri Lankan textile and apparel industry. Government bodies and industry stakeholders can create a supportive environment for sustainable practices and drive positive change in this vital sector by working together. Government agencies can play a pivotal role in setting policies and regulations that incentivize sustainable practices. For instance, they can introduce tax incentives or subsidies for businesses that adopt eco-friendly production methods or adhere to ethical labour standards. Regulations can also establish environmental standards for water usage, emissions, and waste disposal. Collaborative efforts between the government and industry can lead to more effective monitoring and enforcement of sustainability regulations. Regular inspections and audits can ensure that businesses are complying with ethical labour practices and environmental standards, creating a level playing field for all industry players [35, 49].

Government initiatives can support capacity building within the industry. They can provide funding or resources for training programs aimed at enhancing the skills of workers and management in sustainable production and ethical practices. This investment in human capital contributes to long-term sustainability. Collaboration between government research institutions and the industry can drive innovation in sustainable textiles and production methods. Joint research projects can lead to the development of eco-friendly materials, efficient dyeing processes, and resource-saving technologies [61]. Government programs can facilitate access to funding for sustainability initiatives. This can include grants or low-interest loans for businesses investing in more sustainable production methods, such as upgrading machinery to reduce energy consumption or implementing water recycling systems. Government bodies can facilitate stakeholder engagement, bringing together industry associations, non-governmental organizations (NGOs), and academic institutions to address sustainability challenges collectively. Round-table discussions and working groups provide a platform for sharing best practices and finding standard solutions. Government agencies can run public awareness campaigns to educate consumers about sustainable choices and ethical considerations in clothing purchases. This aligns consumer expectations with industry practices and creates demand for sustainable products. Collaboration can also focus on improving market access for sustainable Sri Lankan textiles and apparel. Government and industry can work together to negotiate trade agreements and certifications that recognize the industry's commitment to sustainability, making it easier to access international markets [1, 42].

Circular fashion initiatives are a transformative approach to improving sustainable development in the Sri Lankan textile and apparel industry. These initiatives focus on reducing waste, promoting responsible consumption, and extending the lifespan of clothing items, all of which contribute to a more sustainable and ethical industry. Circular fashion initiatives often include clothing rental services where consumers can borrow garments for a specific period. This approach reduces the demand for constant new production and encourages consumers to wear various styles without purchasing new items regularly. It's a win–win for consumers and the environment. Encouraging second-hand shopping is a key component of circular fashion. Thrift stores, online resale platforms, and clothing swaps allow consumers to buy and sell pre-owned clothing. This extends the life of garments and diverts clothing from landfills, reducing textile waste. Circular fashion emphasizes repairing and altering clothing rather than discarding it when minor issues arise [33, 47]. Businesses can offer repair services or partner with local tailors to make it easier for consumers to mend their garments and keep them in use for longer. Upcycling involves transforming old or discarded clothing items into new and unique pieces. Remanufacturing consists of disassembling and reconstructing garments to restore them to like-new condition. These practices reduce waste and offer creative solutions for fashion enthusiasts. Circular fashion encourages clothing design with durability in mind. This means using high-quality materials and construction techniques that ensure garments last longer. Businesses can create designs that are timeless, versatile, and less likely to go out of style quickly. Educating consumers about circular fashion principles is essential. Businesses can use their platforms to inform consumers about the benefits of second-hand shopping, clothing care, and responsible disposal. Consumer awareness drives demand for circular fashion and fosters responsible consumption habits. Collaborations between apparel brands and second-hand platforms can extend the lifecycle of clothing items. Brands can encourage consumers to sell gently used clothing through these platforms, promoting a circular approach and reducing the industry's environmental impact. Circular fashion initiatives can advocate for sustainability policies like extended producer responsibility (EPR) programs. These policies require manufacturers to take responsibility for the end-of-life management of their products, incentivizing them to design for durability and recyclability [27, 40].

4 Conclusion

In conclusion, sustainable development approaches are not just a choice but necessary for the Sri Lankan textile and apparel industry. This vital sector plays a significant role in the country's economy, but it also faces critical challenges related to environmental sustainability, ethical labour practices, and consumer expectations. By embracing sustainable development approaches, the industry can address these challenges while positioning itself for long-term success and competitiveness. The Sri Lankan textile and apparel industry has the opportunity to lead by example, demonstrating that sustainability and economic prosperity are not mutually exclusive. By embracing these sustainable development approaches, the industry can positively impact the environment, society, and its long-term resilience. It can also meet the evolving demands of consumers who increasingly seek responsible and ethical fashion choices. In doing so, Sri Lanka's textile and apparel industry can secure a sustainable and competitive future in the global market while contributing to a more sustainable world.

References

1. Aigbedo H (2021) An empirical analysis of the effect of financial performance on environmental performance of companies in global supply chains. J Clean Prod 278:121741. https://doi.org/10.1016/j.jclepro.2020.121741
2. Allwood J, Laursen S, Rodriguez C, Bocken N (2006) Well dressed? In the University of Cambridge Institute for Manufacturing Mill Lane. http://www.ifm.eng.cam.ac.uk/uploads/Resources/Other_Reports/UK_textiles.pdf
3. Amarakoon U, Weerawardena J, Verreynne ML (2018) Learning capabilities, human resource management innovation and competitive advantage. Int J Hum Resour Manag 29(10):1736–1766. https://doi.org/10.1080/09585192.2016.1209228
4. Amui LBL, Jabbour CJC, de Sousa Jabbour ABL, Kannan D (2017) Sustainability as a dynamic organizational capability: a systematic review and a future agenda toward a sustainable transition. J Clean Prod 142:308–322. https://doi.org/10.1016/j.jclepro.2016.07.103
5. Annunziata E, Pucci T, Frey M, Zanni L (2018) The role of organizational capabilities in attaining corporate sustainability practices and economic performance: evidence from Italian wine industry. J Clean Prod 171:1300–1311. https://doi.org/10.1016/j.jclepro.2017.10.035
6. Aslam H, Blome C, Roscoe S, Azhar TM (2018) Dynamic supply chain capabilities: how market sensing, supply chain agility and adaptability affect supply chain ambidexterity. Int J Oper Prod Manag 38(12):2266–2285. https://doi.org/10.1108/IJOPM-09-2017-0555
7. Beske P (2012) Dynamic capabilities and sustainable supply chain management. Int J Phys Distrib Logist Manag 42(4):372–387. https://doi.org/10.1108/09600031211231344
8. Bettini Y, Brown RR, de Haan FJ (2015) Exploring institutional adaptive capacity in practice: examining water governance adaptation in Australia. Ecol Soc 20(1). https://doi.org/10.5751/ES-07291-200147
9. Bour KB, Asafo AJ, Kwarteng BO (2019) Study on the effects of sustainability practices on the growth of manufacturing companies in urban Ghana. Heliyon 5(6):e01903. https://doi.org/10.1016/j.heliyon.2019.e01903
10. Brandenburg M, Gruchmann T, Oelze N (2019) Sustainable supply chain management—a conceptual framework and future research perspectives. Sustainability 11(24). https://doi.org/10.3390/SU11247239
11. Caniato F, Caridi M, Crippa L, Moretto A (2012) Environmental sustainability in fashion supply chains: an exploratory case-based research. Int J Prod Econ 135(2):659–670. https://doi.org/10.1016/j.ijpe.2011.06.001
12. Cantele S, Zardini A (2018) Is sustainability a competitive advantage for small businesses? An empirical analysis of possible mediators in the sustainability–financial performance relationship. J Clean Prod 182:166–176. https://doi.org/10.1016/j.jclepro.2018.02.016
13. Ceptureanu EG, Ceptureanu SI, Bologa R, Bologa R (2018) Impact of competitive capabilities on sustainable manufacturing applications in Romanian SMEs from the textile industry. Sustainability 10(4). https://doi.org/10.3390/su10040942

14. Chege SM, Wang D (2020) The influence of technology innovation on SME performance through environmental sustainability practices in Kenya. Technol Soc 60:101210. https://doi.org/10.1016/j.techsoc.2019.101210
15. Choi SB, Feng Y, Liu J, Zhu Q (2019) Motivating corporate social responsibility practices under customer pressure among small- and medium-sized suppliers in China: the role of dynamic capabilities. Corp Soc Responsib Environ Manag 26(1):213–226. https://doi.org/10.1002/csr.1673
16. Choi Y, Yu Y (2014) The influence of perceived corporate sustainability practices on employees and organizational performance. Sustainability 6(1):348–364. https://doi.org/10.3390/su6010348
17. Ciccullo F, Pero M, Gosling J, Caridi M, Purvis L (2020) When sustainability becomes an order winner: linking supply uncertainty and sustainable supply chain strategies. Sustainability 12(15). https://doi.org/10.3390/su12156009
18. Clarke-Sather A, Cobb K (2019) Onshoring fashion: worker sustainability impacts of global and local apparel production. J Clean Prod 208:1206–1218. https://doi.org/10.1016/j.jclepro.2018.09.073
19. de Moura GB, Saroli LG (2021) Sustainable value chain management based on dynamic capabilities in small and medium-sized enterprises (SMEs). Int J Logist Manag 32(1):168–189. https://doi.org/10.1108/IJLM-01-2020-0044
20. del Mar Alonso-Almeida M, Buil-Fabregà M, Bagur-Femenías L, Aznar-Alarcón JP (2017) Shedding light on sustainable development and stakeholder engagement: the role of individual dynamic capabilities. Sustain Dev 25(6):625–638. https://doi.org/10.1002/sd.1682
21. Dhiman R, Sharma M (2017a) Export competitiveness of Indian textile industry: revealed comparative advantage analysis. Int J Appl Bus Econ Res 15(9):295–305
22. Dhiman R, Sharma M (2017b) Productivity trends and determinants of Indian textile industry: a disaggregated analysis. Int J Appl Bus Econ Res 15(22):113–124
23. Dhiman R, Sharma M (2019) Relation between labour productivity and export competitiveness of indian textile industry: co-integration and causality approach. Vision 23(1):22–30. https://doi.org/10.1177/0972262918821230
24. Dhiman R, Kumar V, Rana S (2020) Why export competitiveness differs within Indian textile industry? Determinants and empirical evidence. Rev Int Bus Strat 30(3):375–397. https://doi.org/10.1108/RIBS-03-2020-0021
25. Dubey R, Gunasekaran A, Childe SJ, Papadopoulos T, Helo P (2019) Supplier relationship management for circular economy: influence of external pressures and top management commitment. Manag Decis 57(4):767–790. https://doi.org/10.1108/MD-04-2018-0396
26. Eslami Y, Dassisti M, Lezoche M, Panetto H (2019) A survey on sustainability in manufacturing organisations: dimensions and future insights. Int J Prod Res 57(15–16):5194–5214. https://doi.org/10.1080/00207543.2018.1544723
27. E-Vahdati S, Zulkifli N, Zakaria Z (2019) Corporate governance integration with sustainability: a systematic literature review. Corp Gov 19(2):255–269. https://doi.org/10.1108/CG-03-2018-0111
28. Fiksel J, Bruins R, Gatchett A, Gilliland A, Ten Brink M (2014) The triple value model: a systems approach to sustainable solutions. Clean Technol Environ Policy 16(4):691–702. https://doi.org/10.1007/s10098-013-0696-1
29. Gabler CB, Richey RG, Rapp A (2015) Developing an eco-capability through environmental orientation and organizational innovativeness. Ind Mark Manage 45(1):151–161. https://doi.org/10.1016/j.indmarman.2015.02.014
30. Gam HJ, Banning J (2020) Teaching sustainability in fashion design courses through a zero-waste design project. Cloth Text Res J 38(3):151–165. https://doi.org/10.1177/0887302X20906470
31. Gardetti MA, Senthilkannan Muthu S (eds) (2020) The UN sustainable development goals for the textile and fashion industry. August 2019, 81. http://www.springer.com/series/13111
32. Gbolarumi FT, Wong KY, Olohunde ST (2021) Sustainability assessment in the textile and apparel industry: a review of recent studies. IOP Conf Ser Mater Sci Eng 1051(1):012099. https://doi.org/10.1088/1757-899x/1051/1/012099

33. Govindan K, Shaw M, Majumdar A (2021) Social sustainability tensions in multi-tier supply chain: a systematic literature review towards conceptual framework development. J Clean Prod 279:123075. https://doi.org/10.1016/j.jclepro.2020.123075
34. Hahn T, Pinkse J, Preuss L, Figge F (2015) Tensions in corporate sustainability: towards an integrative framework. J Bus Ethics 127(2):297–316. https://doi.org/10.1007/s10551-014-2047-5
35. Hashim M, Nazam M, Abrar M, Hussain Z, Nazim M, Shabbir R (2021) Unlocking the sustainable production indicators: a novel TESCO based fuzzy AHP approach. Cogn Bus Manage 8(1). https://doi.org/10.1080/23311975.2020.1870807
36. Hofmann KH, Theyel G, Wood CH (2012) Identifying firm capabilities as drivers of environmental management and sustainability practices—evidence from small and medium-sized manufacturers. Bus Strateg Environ 21(8):530–545. https://doi.org/10.1002/bse.739
37. Jha MK, Rangarajan K (2020) Analysis of corporate sustainability performance and corporate financial performance causal linkage in the Indian context. Asian J Sustain Soc Responsibil 5(1). https://doi.org/10.1186/s41180-020-00038-z
38. Jia F, Yin S, Chen L, Chen X (2020) The circular economy in the textile and apparel industry: a systematic literature review. J Clean Prod 259:120728. https://doi.org/10.1016/j.jclepro.2020.120728
39. Kozlowski A, Searcy C, Bardecki M (2015) Corporate sustainability reporting in the apparel industry an analysis of indicators disclosed. Int J Product Perform Manag 64(3):377–397. https://doi.org/10.1108/IJPPM-10-2014-0152
40. Krechovská M, Procházková PT (2014) Sustainability and its integration into corporate governance focusing on corporate performance management and reporting. Procedia Eng 69:1144–1151. https://doi.org/10.1016/j.proeng.2014.03.103
41. Kumar A, Moktadir MA, Khan SAR, Garza-Reyes JA, Tyagi M, Kazançoğlu Y (2020) WITHDRAWN: behavioural factors on the adoption of sustainable supply chain practices. Resour Conserv Recycl X:100031. https://doi.org/10.1016/j.rcrx.2020.100031
42. Latan H, Chiappetta Jabbour CJ, de Sousa L, Jabbour AB, Wamba SF, Shahbaz M (2018) Effects of environmental strategy, environmental uncertainty and top management's commitment on corporate environmental performance: the role of environmental management accounting. J Clean Prod 180:297–306. https://doi.org/10.1016/j.jclepro.2018.01.106
43. Libera A, de Barros FPJ, Faybishenko B, Eddy-Dilek C, Denham M, Lipnikov K, Moulton D, Maco B, Wainwright H (2019) Climate change impact on residual contaminants under sustainable remediation. J Contam Hydrol 226:1–42. https://doi.org/10.1016/j.jconhyd.2019.103518
44. Liu L, Zhang M, Ye W (2019) The adoption of sustainable practices: a supplier's perspective. J Environ Manage 232:692–701. https://doi.org/10.1016/j.jenvman.2018.11.067
45. MacCarthy BL, Jayarathne PGSA (2012) Sustainable collaborative supply networks in the international clothing industry: a comparative analysis of two retailers. Prod Plann Control 23(4):252–268. https://doi.org/10.1080/09537287.2011.627655
46. Macchion L, Moretto A, Caniato F, Caridi M, Danese P, Spina G, Vinelli A (2017) Improving innovation performance through environmental practices in the fashion industry: the moderating effect of internationalisation and the influence of collaboration. Prod Plann Control 28(3):190–201. https://doi.org/10.1080/09537287.2016.1233361
47. Mandarić D, Hunjet A, Vuković D (2022) The impact of fashion brand sustainability on consumer purchasing decisions. J Risk Fin Manage 15(4). https://doi.org/10.3390/jrfm15040176
48. Moreira Mero N, Hidalgo Fernández A, Loor Alcívar MI, González Santa Cruz F (2020) Influence of internal marketing dimensions on organizational commitment: an empirical application in ecuadorian co-operativism. SAGE Open 10(3). https://doi.org/10.1177/2158244020945712
49. Muhammad N, Scrimgeour F, Reddy K, Abidin S (2015) The relationship between environmental performance and financial performance in periods of growth and contraction: evidence from Australian publicly listed companies. J Clean Prod 102:324–332. https://doi.org/10.1016/j.jclepro.2015.04.039

50. Nazam M, Hashim M, Nută FM, Yao L, Zia MA, Malik MY, Usman M, Dimen L (2022) Devising a mechanism for analyzing the barriers of blockchain adoption in the textile supply chain: a sustainable business perspective. Sustainability 14(23). https://doi.org/10.3390/su142316159
51. Olawumi TO, Chan DWM (2020) Concomitant impediments to the implementation of smart sustainable practices in the built environment. Sustainable Prod Consump 21:239–251. https://doi.org/10.1016/j.spc.2019.09.001
52. Oleinykova O, Panchenko M, Tymko Y (2019) Strategic approaches to business diversification. Econ Fin Law 12/1:28–30. https://doi.org/10.37634/efp.2019.12(1).6
53. Paul J, Dhiman R (2021) Three decades of export competitiveness literature: systematic review, synthesis and future research agenda. Int Mark Rev 38(5):1082–1111. https://doi.org/10.1108/IMR-12-2020-0295
54. Pham H, Kim SY (2019) The effects of sustainable practices and managers' leadership competences on sustainability performance of construction firms. Sustain Prod Consump 20:1–14. https://doi.org/10.1016/j.spc.2019.05.003
55. Rahman O, Hu D, Fung BCM (2023) A systematic literature review of fashion, sustainability, and consumption using a mixed methods approach. Sustainability 15(16). https://doi.org/10.3390/su151612213
56. Rana S (2019) Sustainability in business: some research perspectives. FIIB Bus Rev 8(2):77–78. https://doi.org/10.1177/2319714519854232
57. Redman CL (2014) Linked references are available on JSTOR for this article : should sustainability and resilience be combined or remain distinct pursuits? Ecol Soc 19(2):37. http://www.ecologyandsociety.org/vol19/iss2/art37/
58. Sarfraz M, Ivascu L, Belu R, Artene A (2021) Accentuating the interconnection between business sustainability and organizational performance in the context of the circular economy: the moderating role of organizational competitiveness. Bus Strateg Environ 30(4):2108–2118. https://doi.org/10.1002/bse.2735
59. Selimović Š, Khademhosseini A (2011) Research highlights. Lab Chip 11(18):3029–3030. https://doi.org/10.1039/c1lc90080c
60. Shen B, Li Q, Dong C, Perry P (2017) Sustainability issues in textile and apparel supply chains. Sustainability 9(9):1–6. https://doi.org/10.3390/su9091592
61. Sivagnanasundaram M (2018) Sustainability practices in indian retail industry: a comparison with top global retailers. Emerg Econ Stud 4(1):102–111. https://doi.org/10.1177/2394901518770025
62. Souza AAA, Alves MFR, Macini N, Cezarino LO, Liboni LB (2017) Resilience for sustainability as an eco-capability. Int J Clim Change Strat Manage 9(5):581–599. https://doi.org/10.1108/IJCCSM-09-2016-0144
63. Sudusinghe JI, Seuring S (2020) Social sustainability empowering the economic sustainability in the global apparel supply chain. Sustainability 12(7):1–18. https://doi.org/10.3390/su12072595
64. Sze N, Brocke J (2017) What can we learn from corporate sustainability reporting? Deriving propositions for research and practice from over 9, 500 corporate sustainability reports published between 1999 and 2015 using topic modelling technique. PLoS One: 1–27
65. Teece DJ (2012) Dynamic capabilities: routines versus entrepreneurial action. J Manage Stud 49(8):1395–1401. https://doi.org/10.1111/j.1467-6486.2012.01080.x
66. Tseng ML, Chang CH, Lin CWR, Wu KJ, Chen Q, Xia L, Xue B (2020) Future trends and guidance for the triple bottom line and sustainability: a data driven bibliometric analysis. Environ Sci Pollut Res 27(27):33543–33567. https://doi.org/10.1007/s11356-020-09284-0
67. Türkay M, Saraçoğlu Ö, Arslan MC (2016) Sustainability in supply chain management: aggregate planning from sustainability perspective. PLoS ONE 11(1):1–18. https://doi.org/10.1371/journal.pone.0147502
68. Weerasinghe N, Weerasinghe A, Perera Y, Tennakoon S, Rathnayake N, Jayasinghe P (2023) Sustainability practices and organizational performance during the COVID-19 pandemic and economic crisis: a case of apparel and textile industry in Sri Lanka. PloS One 18(7). https://doi.org/10.1371/journal.pone.0288179

69. Yussif A-M, Belko S, Oavare OP (2022) CSR as an Elixir for enhanced corporate image: experiences from the university for development studies, tamale and C.K. Tedam University of Technology and Applied Sciences, Navrongo, Ghana. Int J Human Soc Stud 10(6):54–61. https://doi.org/10.24940/theijhss/2022/v10/i6/hs2204-020

"The Quality that Comes from Oral Tradition is the Trend and Luxury." Indigenous Hand-Loom Textiles Show Potential Towards Sustainable Development

G. M. Ranathunga and A. M. Lakshani Imeshika

Abstract This chapter highlights that oral tradition holds the identity of indigenous cultural textiles, which shows that the quality that comes from tradition is the trend and luxury and gives insights towards sustainable development. The village of 'Thalagune' in the Central hills of Sri Lanka has earned a long reputation for identical hand-loom textiles since ancient times. Today, 'Thalagune' textile productions are known as one of the best commercial cultural products among locals and foreigners. They are a community with 8 families living in an interconnected society and selling their skills. The research found that indigenous weavers have improved their quality of life by understanding and realising their potential in the art and craft of traditional hand-loom weaving. The way of life contains the capacity for continuance, which is bound with the inheritance memory. Folk poems/verses, stories, ritual recites, and beliefs show their valuing system of life and occupation. Oral tradition reveals the whole product development procedure. From cultivation of raw materials, preparation of cotton yarn, dyeing, feeding to weaving machines to weaving patterns, traditional design and motifs and a variety of products were recorded in their oral tradition. Many factors cause inequalities in the textile industry, such as gender wage, trade liberation, technology, lack of competent labour, and erratic yarn supply. The inability to make a livelihood from one's skill also causes inequality in the textile industry. Besides, Luxury brands look for identity and uniqueness. The process of design in traditional hand-loom weaving shows the potential to provide firm alternatives. Traditional weaving shows unique aesthetic expression as they have been inherited from generation to generation. To strengthen the livelihood from one's skill, it is suggested to develop local perspectives, have the service of traditional artisans' skills, and let them earn. When designers join the instinctive with the mechanical, the hand-crafted and the technological, the result can be magical. This chapter found its close relation with sustainable development goal 10, which reduced Inequalities:

G. M. Ranathunga (✉) · A. M. Lakshani Imeshika
Department of Textile and Apparel Engineering, University of Moratuwa, Moratuwa, Sri Lanka
e-mail: gayathrir@uom.lk

A. M. Lakshani Imeshika
e-mail: a.m.lakshani@gmail.com

G. M. Ranathunga (ed.), *Textiles of Sri Lanka*, SDGs and Textiles,
https://doi.org/10.1007/978-981-97-0569-6_2

Getting a better life and well-being would help reduce inequalities in the textile industry.

Keywords Oral tradition · Memory · Intangible heritage · Hand-loom textiles · Identity · Well-being · Sustainable survival · Sustainable development

1 Introduction

SDG Goal 10: Getting a Better Life and Well-Being Would Help Reducing Inequalities in the Textile Industry

Many factors cause inequalities in the textile industry. Gender wage [9], trade liberation and pay [21], and technology [24] are a few elements that affect inequality. Further, the sector encounters challenge due to a lack of competent labour [32], p. 2 and an inconsistent yarn supply. The difficulty of creating or accessing tools, the unavailability of the raw material, or the lack of adequate substitute materials are all challenges, the inability to make a livelihood from one's skill is also a challenge [30], p. 7. Since money is always needed, the uneven financial distribution is another element that affects the next generation's interests. Lack of technology and understanding that innovative designs are rapidly becoming outdated is a fundamental problem facing traditional hand-loom cooperative societies [32], p. 8. To solve problems, specific cultural art and craft industries have developed regional models. According to [10], p. 689, India's hand-loom sector may manufacture goods that are marketable in small quantities, have room for innovation, switch to the newest designs in line with market trends, be adaptable to meet supplier's requirements, and develop great plans. To empower hand-loom workers to operate independently and increase their income levels, empowerment policies must educate them, offer credit facilities, give yarn at discounted prices, and integrate modern marketing methods [10], p. 700, [32], p. 2.

Issues and solutions are at different levels and address local perspectives. Besides, the process of design in hand-loom weaving shows the potential to provide firm alternatives. The global fashion industry's economics is based on competition, with consumption being charged by low costs obtained through mass production. Meanwhile, expensive multinational brands are available, and their ambitious customers embrace the consumption of high-end goods [3, 16], p. 4, [30], p. 5. These worldwide fashion firms' customers actively pursue high living standards. Luxury brands are pricey because of the high-tech, expensive materials, product development procedures, premium designs, and designer involvement. Numerous nations use national models and recommend value additions for the luxury market. According to [30], p. 5, to give new interpretations to luxury branded clothing, the world must view India's craft traditions through unique views considering the current global fashion crisis or the luxury market of fashion brands. India can provide a traditional artisanal alternative that is adaptable, sustainable, and genuinely abundant at the higher end of the quality scale. Luxury brands may find the most sought-after identity, and uniqueness from different perspectives, in the meantime traditional artisans could earn more from their inherited values which is a win–win situation, giving a successful system to reduce the inequality in the textile industry. The show Manus × Machina: Fashion in an Age of Technology, which featured garments manufactured both by hand and by machine, showed that the heart must act as a bridge between the human hand and the machine [39], pp. 257–263. Both hand-crafted and machine-produced goods must be made with equal care and attention to quality. The hand and the machine are "equal and mutual protagonists in solving design problems," not "oppositional". Andrews Bolton stressed that "I'm very wary of becoming too reliant on technology because the design process has to remain instinctive." What does seem clear in the exhibit is that when designers join the instinctive with the mechanical, the hand-crafted and the technological, the result can indeed be magical. Dutch designer Van Herpen expresses "I work with technology, but the hand and machine are equal in my design process, they are integrated" [39], pp. 257–263.

The idea offers a fresh perspective on Sri Lanka's textile sector. The hand-loom textile industry of Sri Lanka shows high calibre in design and product development, driven by an age-old traditional art and craft system [20, 38]. Traditional weaving in traditional society shows the well-being of life. Identity, enthusiasm, and self-esteem create a unique art which affects the well-being of the traditional art and the weaver in the hand-loom industry. The selected study setting is Thalagune, a traditional hand-loom weaving village in Kandy District in the Central Province of Sri Lanka. This community is the oldest village for traditional hand-loom weaving in Sri Lanka. Dumbara weaving patterns are recognised as a cultural heritage. The chronicle Mahavamsa mentioned that Sri Lanka has greatness for spinning and weaving and takes pride in producing high-quality cotton fabric. Dumbara weaving tradition is the sole remaining example of a traditional weaving system where native weavers still use poems, songs, and stories to guide their work and remind them of the proper practices for generations. From raw material supply to production, artisans recite the traditional process orally. A humble verse or poem holds a nation's intangible heritage. Weaving techniques, design and motifs, design layouts, tools, raw materials, product development process, machinery, careers, way of life, religion, culture, traditional society, norms, and values described in poems/verses are the same as their inheritance used in the past. Poems act as a value-retaining system of the community's well-being, art, and occupation. It is an educational system where all values are preserved, serving as a resource pool for the future.

2 Literature Review

2.1 What is the Well-Being of a Person?

Good physical, mental, emotional, and social health is well-being [25]. It serves as a source of meaning and purpose for daily life in addition to being just a feeling or mood. The individual's well-being is what they value most in life and what is best for them [37]. As a result, in traditional culture, the health of the arts and crafts and the population's "quality of life" impact the individual, the community, and the economy [18]. Artisans care for art and craft, and their active participation in art and skill will benefit their well-being. Consequently, the connection between craft activities, creative engagement, mental and physical stimulation, social contact, and self-esteem contributes to the well-being of the artisan and the craft itself [7, 14, p. 3, [36], p. 508, [1], p. 54. Depending on the culture, region, and gender, different crafts have different functions and relationships with their creators. The fact that crafters are frequently emotionally committed in their pursuits, and many persist through all phases of life and into old age, is a common thread, though.

2.2 Culture, Tradition, and Sustainable Rural Life Create a Traditional Art and Craft Identity

Rural communities can survive and prosper in traditional or vernacular cultures [4], p. 499. In traditional civilisation, survival and sustainability were intertwined. The sustainability of the art and craft of a traditional community depends on the maintenance of resources [15], p. 3441. Their culture offers a common destiny based on shared practices, memories, and values. It aids in mobilising and disseminating survival techniques among community members, linking their traditional skills and traditions to their way of life. The villagers are aware that their conventional jobs continue to be their main source of income and that leaving them would be difficult [4], p. 504. Culture, tradition, and sustainable rural life create a traditional art and craft identity. Edward [31], p. 11 states "tradition is created through human activities, thoughtful consideration, and creativity. Therefore, tradition refers to everything passed down from previous generations to the present". Tradition and culture are intertwined. Coomaraswamy [2] considered art an instrument through which tradition is expressed. According to [31], p. 6 the greatest treasure in each society is its traditional knowledge. The cultural experience shapes, develops, modifies, and transforms the knowledge of tradition. A culture stands upon mutual interrelations of state and religion. Culture affects concurrent political, administrative, social, and religious factors. The political constitution is the main factor in navigating a country's cultural perception [26], p. 23. Cultural values that have a lasting and independent influence on society are essential to the long-term survival and prosperity of an ancient traditional artisanal civilisation. According to [30], p. 9, crafts are lived practices that constantly change in response to societal, economic, and cultural demands. The region's activities, such as the modernisation of lifestyles and the transformation of manufacturing, will also alter. Residents demand and expect these improvements because they provide them with a sense of security and belonging and contribute to creating a sustainable and healthy environment [42], p. 24. Traditional crafts, arts, and ways of life have a survival pattern. They enjoy weaving attractive clothes [6]. Development and well-being are intertwined, according to [19], p. ii, who proposed a holistic understanding of Maori cultural creation in New Zealand. Cultural assets significantly impact economic outcomes, cultural retention and use, the application of Maori values, and well-being. Preserving natural resources that are necessary for survival is regarded as being part of sustainability. It focuses on the conflict between humankind's desires for a better existence and nature's constraints on the other side [15]. McFarlane [19] claims that enhancing and maintaining cultural assets helps people develop their capacities (human assets), maintain their relationships with their environment (natural assets), and give them a sense of identity, self-worth, and dignity. It also improves social assets, including interpersonal connections and trust.

2.3 The Value-Retaining System of the Traditional Societies: Oral, Another Name for Tradition, is the Survival Power of Inheritance

Traditional knowledge is passed orally for generations from one person to another within a particular community. It can be expressed through folklore, rituals, songs, stories, legends or customs [43], p. 23. Oral means 'spoken',' verbal'. 'unwritten'. Transmission gives the meaning of handed down [34]. p. 195. Oral tradition focuses on recollections of the past that extend beyond one's lifetime. They are verbal communications that report on the past as recalled by those in the present. While they are personal memories, they are someone else's memories—a parent, a grandparent, an elder—passed down through oral statements, sung, or played on musical instruments. These traditions are part of an intangible heritage that cannot be captured in any other way since those who had first-hand knowledge about or participated in a historical event or way of life are no longer living' [17], p. 19. Oral history is foremost in their historical minds [17], p. 21. 'Oral provides clues not only about the meanings of historical experience but also about the relationships between the past and the present, between memory and personal identity, and between individual and collective memory, relationships that the state with its traditional, official narratives has only recently started to embrace' [17], p. 22.

"Oral history enables people to think more deeply about their place," claims [17], p. 22. Poems depict crafts people's way of life and environment [8], p. 8. The study of the relationship between man and land has centred on ideas like place and placeness. Placeness is simply the meaning that people give to a space, and it refers to a place's singularity within a particular time and space [42], 2p. According to Daskon and Binns' analysis [4], p. 508, "livelihoods are geographically and culturally predetermined, and the inherent skills are disseminated among the community members, not only for their economic survival but also to ensure their identity within the community." The examination of traditional Thai textile weaving by [40], p. 56 confirms that the product identity is owned by a particular community in particular geographical data and information with distinctive attributes of the product.

Folklore and literature possess a strong sense of survival [34]. Oral, transmission, tradition, survival, and communal qualities are connected to the survival of legends

[34], p. 195. Fong [8] noted how women's poetry is intertwined with everyday activities in traditional Chinese embroidery. The expertise of embroidering was passed down directly from mother to daughter, aunt to niece, older sister to younger sister, and woman to woman because it was so closely related to daily living in the inner quarters. It was inextricably linked to women's perception of the ordinary [8], p. 66. Folklore is another synonym for tradition because verbal communication is important [34], p. 193. Everything is folklore in an entirely oral civilisation [34], p. 194. Folklore generates original concepts that have developed into cultural norms among members of many civilisations and can be identified as their shared characteristics [34], p. 195. Utley [34], p. 196 states that customs, beliefs, and rituals generate folklore, proverbs, and riddles. It can be defined when an idea or superstition is presented in a more refined aesthetic framework. The story is created and kept in the knowledgeable teller's memory.

In the nineteenth century, development became necessary. Traditional societies lacked developmental techniques. Traditional society's rules, values, and customs make up this society. According to [19], p. ii, indigenous communities are subjected to imposed development, where "success" is determined by achieving economic goals. In traditional civilisation, sustainability and development go hand in hand. According to Max Weber and Samuel Huntington [13], p. 19, cultural values have a long-lasting and independent impact on society. Planning for development should consider the values, meanings, customs, and knowledge systems that people and communities have that affirm identity and variety. Daskon and Binns [4], p. 495 suggest "The knowledge, customs, capabilities, and ingenuities which can play an invaluable role in sustainable community development". Seeing the integral role of culture in community sustainability is made possible [23]. Perception of cultural capital is described in three forms: embodied (e.g., values, norms and attitudes, knowledge), objectified (cultural goods, e.g., crafts, architecture, instruments), and institutionalised (educational credentials) [4], p. 511. The functional aspect aids in comprehending "what culture serves" the community, including identity formation, coping mechanisms, and survival abilities [4], p. 511. Coomaraswamy [2] saw tradition as a powerful tool for education. The artist, the patron, the consumer of it, one and all are shaped and invigorated by a specific tradition. Since culture is related to money, it is easily accessible as a capital or asset. Further, rational, tolerant, trusting, and participatory attitudes progressively replace absolute standards and ideals, economic development also correlates with these changes. It is important to carefully identify the transferable characteristics, a value-retaining system of art and a craft-based economic constitution.

3 Methodology

The research is a qualitative research approach to investigate how oral tradition retains the identity of indigenous cultural textiles, which shows that the quality that comes from tradition is the trend and luxury. Oral histories can be used to better understand broader social and cultural history in ways that literature, data, and evidence cannot. From an anthropological perspective, oral history is a method of qualitative interviewing that emphasises participants' views and the experiential knowledge of people living through the events [17], p. 20. At its best, the interview is an adynamic, dialogical relationship that encourages active remembering and meaning-making [17], p. 20).

The selected study setting is Thalagune textile village in Kandy District, Central province, Sri Lanka (Fig. 1). Data obtained from secondary and primary sources. The research commenced with a survey of literature from secondary sources. The literature review concentrated on one important issue: How oral tradition acts and serves in society to continue the longevity and well-being. Primary sources were collected from participant observation. Participant observation was carried out to gather information from folk poems, stories, ritual recites, and beliefs of their daily routine. Folk poems of 12, stories, ritual recites, and beliefs related to Dumbara textile weaving tradition were collected. A systematic oral record of legends reveals raw material preparation—cotton, tools and production process, weaving patterns, design and motifs (flower types, animal designs, geometric shapes), and product categories. In this study, descriptive interviews were used to collect data. To collect details on the subject, 8 interviews were conducted throughout three rounds (Table 1). One pilot interview and two in-detail interviews were held through participant observation. Some clarifications were carried out through telephone conversations. "Interviews are descriptive and allow the interviewer to explore ideas freely" [28]. Interviews are divided into occupation categories: weavers, raw material producers, and producers of traditional weaving machines. Questions are structured to get more in-depth responses. Interviews were held, including weavers of different ages and experience levels. Oral data were recorded to an audio recorder and then transcribed. Transcribed data were observed and categorised according to the product development procedure.

Data analysis commenced from the observational study. Using a word processing tool, the recorded interviews were transcribed and coded. Poems or folk stories were categorised according to themes such as legendary memories of traditional weaving of Thalagune village. Gathered data was coded and the production procedure was identified step by step. Folk poems/verses are categorised according to sub-topics of the production procedure.

Fig. 1 Aerial map of Thalagune village. Central province of Sri Lanka

4 Artisans' Potential in the Art and Craft of Traditional Hand-Loom Weaving

The poem reveals a circle of systems which facilitate one another's survival. Traditional knowledge from raw material findings or cultivation, harvesting, preparation of raw material, production, techniques, tools, design and motifs, colour, range of productions, market and definite customers is retained in poems.

Table 1 Details of the interviews

Number of the interviewee	Interviewee details	Purpose of the interview
Interviewee 01	Weaver	Colour mixing
Interviewee 02	Weaver	Legendary stories related to weaving Product range—ancient and contemporary Variety of patterns Design and motifs Machine types and tools
Interviewee 03	Weaver	Contemporary patterns
Interviewee 04	Weaver	Variety of patterns
Interviewee 05	Weaver	Legendary stories related to weaving
Interviewee 06	Weaver	The oldest notion of patterns Colours—contemporary Traditional weaving machines
Interviewee 07	Weaver	Legendary stories related to weaving Variety of patterns Design and motifs
Interviewee 08	Weaver	Colours—ancient and contemporary Traditional machines

4.1 Legendary Memories of Their Weaving History, Lineage, and Village Are Their Biggest Resource

Historical legends about weaving in Thalagune village go back to the country's earliest recorded history. According to folk stories of the village, the origins of Dumbara weaving go back to Kuweni, a fierce warrior queen who ruled the country in the fifth century BC and Ravana, the powerful mighty King in ancient epics. Prince Vijaya from India met Kuweni while spinning cotton threads; after she was expelled; she cursed and burned her cotton farm in Dumbara, where ashes rose high in the sky and blanketed the land with patterns. The legend becomes an inspiration for Dumbara weaving patterns. The tradition has been passed down from generation to generation since then. Weavers remember their ancestors' history and remember it with utmost pride.

> Kuveni was winding cotton when King Vijaya arrived in Sri Lanka, according to the chronicle Mahavansa. The strongest evidence that dumbara designs have been passed down through generations is found in Mr. Ananda Coomaraswamy's book, Medieval Sinhala Arts. (Interviewers 2 and 5, Weaver, 2023).

> My grandmother is descended from Kuveni. we are the ninth generation of Dumbara inherited weaving tradition. (Interviewer 5, Weaver, 2023)

> We know epics related to our village. king ravana lived in a nearby area called 'yahangala' since then we have been used to weaving. (Interviewers 2 and 5, Weaver, 2023)

> One of our grandfathers has woven the cloth himself. Those fabrics are available in museums. The king's clothes were woven on one side of the palace. As I remember, my grandfather's

name was Tikira. It is said that he made clothes for King Rajasingha. (Interviewer 2, weaver, 2023)

Coomaraswamy [2], p. 232 states that "the weaving of homespun cotton cloths was once unique in the Kandyan province. It is still alive in the village called Thalagune in 1907". The village's name 'Thalagune' has been derived from a historical story about a king. It is said that the queen gave birth in a hut made of Palmyra leaves in the village (Interviewer 5, Weaver, 2023). Weaver 7 remembers a folk story that the last King of the Kandyan Kingdom hid in the village, and villages treated and took care of the King. Weavers (Interviewers 5,7, weavers, 2023) remember that the sacred tooth relic of the Lord Buddha, in the Temple of the tooth relic, Kandy was once hidden in Thalagune village. The weaver remembers how his ancestors respectfully protected the sacred tooth relic.

Dumbara weaving is given value according to the name and skilful weavers of the village "Thalagune'. Interviewer 2 commented (weaver, 2023), "We value our inheritance". "We must live in this village". "otherwise, no value. Weaving belongs to this village". Rajakariya, a system of caste-based land tenure that allotted businesses according to caste, served as the foundation of Sri Lanka's old governmental structure, which was centred on a monarchy. The King gave service castes land (*nindagam*), and they were given various jobs in exchange. Each caste had economic privilege since it was the only one with the right to provide a specific type of labour [2], p. 232, [5]. The research location, Kandyan art and craft production exhibits a strong connection to Rajakariya's caste structure and land tenure. As interviewees involved in the research point out, the relationship between caste, land and craft industry is essential [35]. It has long been understood that landscape and memory are related [33]. p. 1. "Livelihoods are geographically and culturally predetermined, and the inherent skills are disseminated among the community members, not only for their economic survival but also to ensure their identity within the community" [4], p. 508.

During the Kandyan monarchy era, Kings had given the responsibility to us to produce hand-loom products. The King donated a village to our ancestors, and they respected our lineage. We are the people who produce textiles to the Temple of Tooth Relic. Still, we are in those lands. (Interviewer 7, weaver, 2023)

Storytelling has a long history in Sri Lanka, where jataka stories and anecdotes from the Buddha's life are utilised to convey philosophical ideas [11]. The art and craft of Sri Lanka have been nourished by Buddhist philosophy. Offering the best quality woven textile to the Temple of the Tooth relic is considered as the high merit of their life (Interviewer 2, weaver, 2023). They were often used to visiting the Temple of the Tooth relic, however, the custom has not been practised by all today (Interviewer 1, 2, 4, 5, 6, 7, 8 weavers, 2023). Social customs, norms, rituals have been derived from Buddhist culture. Healing rituals such as 'kohoba kankari', 'labu gediye kankari' are held annually. Folk drama 'sokari' is an essential entertaining event of the villages (Interviewers 1, 2, 3, weavers, 2023). "Customs and religious ceremonies offer the chance to transmit inherited information [12]. Artisans remember that their grandfathers were skilful indigenous physicians who were also competent in witchcraft (Interviewer 3, weaver, 2023).

The villages are relatives and live closely within 1.1/2–2 km. Cultivation and harvesting are group activities called '*aththama*' (Interviewer 1, 2 weavers, 2023). Occasionally, a family gets a big order other weavers help to complete. They live in harmony and mutual understanding. Marriage had previously been strictly among relatives; however, it is seen that such restrictions have faded away over time (Interviewer 6, weaver, 2023). Farming is also associated with customs and rituals such as God offering right after the harvesting of crops (Interviewer 4, weaver, 2023). They still recall poems associated with farming, "Songs, stories, and poems reflect our older generation. But we know some of them. In those days, our ancestors had peaceful, calm lives. Then they sang. Now we don't have time to sing; we are so busy and commercialised (Interviewer 2, 3, Weaver, 2023)". Witchcraft is still popular among people. Ayurvedic medicine is considered as respect and utmost gratitude. Interviewer 3 (Weaver, 2023) remembers their cultural norms and values. Indigenous physicians don't earn money from treatments. People bring only a handful of Beatles for respect.

These weavers learnt this craft from their ancestors. Weavers believe they are responsible for teaching their ancestry knowledge to younger family members. Weavers repeat the same traditional designs. Repetition is an essential tradition characteristic in design development, even when new information is added. Repetition strengthens concepts and improves ideas [29]. Interviewer 2 (Weaver, 2023) states, "My parents taught me all of the Dumbara designs". "Weaving is something we do every day, so through repetition, I can recall all the patterns". This community sings poems to guide weavers through their work and serve as an educative tool for proper practices. "Life and occupation became one" [27]. Interviewer 2 (Weaver, 2023) states, "Poetries exist but are not currently being used. This was carried out while farming during the period of our great grandmother. With the freedom they enjoyed then, they have recited poems.". There are inherited secrets about weaving, which has become a traditional practice. They protect and value their identity and understand its monetary value and transcendent quality.

> We wrap the thread, put it in the warp, and give it to those who come to work with us to weave. We don't want to give away our secrets, so we saved it anyway. We believe the right belongs to our village and family, so we kept it like this. (Interviewer 2, weaver, 2023)

They preserve the weaving techniques that have been passed down to them from their ancestral heritage. Each family has a unique style of colour mixing and patterning, which shows a unique aesthetic in each family. Interviewer 1 (Weaver, 2023) states that "There are numerous weaving families, we have weaving machines at our home as shown in Fig. 2. Each family has a unique method for placing and combining the colours. Everyone has their style. Each of us has a unique individuality." The general method of learning weaving is remembering and prolonged engagement. Ancient weavers did not have literacy ability. Weavers have 8–35 years of expertise achieved from their prolonged engagement in weaving. The younger generation of today have their secondary education. However, all keep the ancient method of remembering designs and weaving techniques (Interviewers 1, 2, 3, 4, 5, 6, 7 weavers, 2023). Innovation is woven right on the loom without trial or sampling (Interviewers 1, 2, 3, 4, 5, 6, 7, weavers, 2023).

> It is impossible to weave just because you have a proper education. you need prolonged engagement. We have an artisan who doesn't know how to measure from measuring tape. The weaver doesn't know numbers. We measure 10 m and give to weave. The weaver is the best. (Interviewer 2, 2023)

> Weaving is our inheritance. We have seen weaving since our childhood, all my designs and techniques are in our memory, we don't need any practice or special education on weaving, if there is a new design we take photograph of it, we can develop weaving technique by looking at an image. (Interviewer 2, weaver, 2023)

They are interested in new trends, learning, and demonstrating their skills with modern knowledge. Weaver 6 said that in 1989, when Sri Lankan architect Tilak Samarawickrama worked with them, they got inspiration from him, which changed the weaving style. The interaction between "the past" and "contemporary practice" creates a fruitful and dynamic cycle. The cultural future is not the same as the cultural past [26]. The meaning of the new era can be presented as "living traditional culture" when culture is subjected to contemporary adaptation and recombination. This can be done by understanding how to incorporate aspects of indigenous legends and

Fig. 2 Weaving machine at Thalagune Village, Central Kandy

historical memory into products and adapt to industrialisation [41]. Tradition changes and then survives. They are adapted to new changes. Interviewer 6 (Weaver, 2023) states, "New trends that are specific to the family have emerged. New colour matching and shape designs are used in older technology. We received new knowledge from Tilak Samarawickrama who was an architect. Using this dumbara pattern technology, we revealed special pattern to the world." (Interviewer 6, weaver, 2023).

- **Raw material Preparation -Cotton farming and harvesting**

> Two individuals worked in two shifts to wind cotton.
> The perfect amount of rain allowed cotton to grow extensively.
> The mountain-dwelling god Shiva promoted extensive cotton growth.
> The emperor Ravana was on his way.
> (Poem 1: a poem of Cotton farming and harvesting, English translation by author)

It is described in the poem one that they engaged in both weaving and agriculture simultaneously. The husband and wife were assigned to work in two shifts. Cotton was grown perfectly since the quarter had received the correct amount of rain. They believe that the god of the mountains, "Shiva" helped ensure the proper rain. The name of the "King Ravana" is mentioned in the poem's final line. Cotton farms were close to the village. As stated in a verse, proper rain is required for healthy productivity. They said that one or two rainfalls were required for a quarter. Cotton takes six to eight months to grow. Raw cotton is twisted to create threads. In the past, cotton was extensively grown on dry fields in Kandyan districts, cotton was usually not spun in the field but taken home and then dried and cleaned in large quantities [2], p. 235. Today, cotton farms are abandoned. They have turned to jungles. Weavers purchase yarns from the market. They mostly use 20/2 cotton yarns. In the past, cotton was coloured using organic dye substances. People can still find trees, plants, and other dye substances in the woodlands close to the village.

- **Weaving Tools and production process**

Dumbara weaving applies patterns to both sides of the fabric, which makes the weaving advanced and complex. The way they use raw materials and how the weave is applied to fabric is unique to these artisans. Poem two describes that by adding thread using a plank between the vertical and horizontal threads of the fabric, the fabric is woven in this tradition.

> The method' dik gaseema' was used to insert yarn.
> The plank is manually pressed for weaving.
> Right hand is used to handle the weaving tool called 'koora patta'.
> The pattern 'thunpath' is used to create the fabric.
> (Poem 2: a poem describes weaving tools and production process/English translation by author)

In ancient times, when artisans inserted warp into machines, they used a method called 'dik gaseema' and then weavers pressed the plank by hand and started weaving

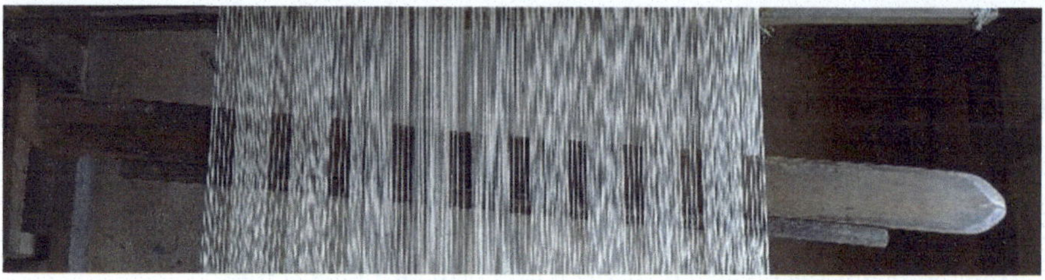

Fig. 3 'Koora patta' tool, which is handled with the right hand, was used to weave

as described in poem two. The 'koora patta' tool, which is handled with the right hand, was an essential tool in Dumbara weaving which is shown in Figs. 2 and 3.

Women who enjoy weaving fascinating clothing.
Women who dislike knotted threads.
Women who weave numerous clothes within thirty hours.
May the gods shield the weavers.
(Poem 3: a poem describes weaving/English translation by author)

The poem three describes that women were engaged in weaving daily. It has been even in the 18th Century in Kandy. "The action of spinning has been charming and graceful; it is women's work, but weaving is done by men and women alike. It is very usual for several women to meet for a spinning bee, when the work is done to accompany all religious songs" [2], p. 235. Culture that is based on Buddhist religion, customs, norms, values are intertwined and facilitate well-being of life [22]. Yarn must be cleaned and free from all kinds of dust. Knots in yarn must be in less number. If yarns have knots, it's difficult to weave. Therefore, women don't like yarns which have knots. Weaving is a descriptive task. It is time consuming. Time, effort, dedication are never counted. They work hard for almost thirty hours. It is said that they don't get enough pay for that. In the last line, they hope all gods protect the weaving women. Positive perception over life is mentioned in the poem.

The rope is tied at the end and weaves.
The plank is pressed by hand and weaves.
'Koora patta' is taken from the right hand and weaves.
The cloth is woven according to the pattern 'pannan'.
(Poem 4: The poem describes how an artisan gradually develops a cloth/English translation by author)

The poem four describes how an artisan gradually develops a cloth. Weaving of the pattern 'Pannan', the rope is tied first and then weaves. Then the plank is pressed by hand. This is the starting point of weaving a cloth. It is seen that both 'pannan' and 'thunpath' patterns show similar processes. Some weavers show high calibre in weaving more complex designs. Weaver 2 proudly remembers the family expertise. Interviewer 2 (Weaver, 2023) states, "Only our 8 Sri Lankan families can create large, complex designs on fabric. We are familiar with the craft. While the cloth is being weaved, design is applied. No matter how big or advanced it is, we are sure to do it."

Cross, flower of cross, knop and flower, flower called 'pana'.
A line of wood pigeons, birds in between them.
Line by line: lion, lion with an elephant's head, elephant, tusker.
Let's bend our necks and weave threads at the' hadawa' weaving machine.
(Poem 5: a poem describes flowers and the type of traditional weaving machine/birds and animals/English translation by author)

The poem five describes a type of traditional tool called 'hadawa'. The oldest weaving machine was called' aluwa'. 'hadawa' was a tool in that machine to tie the yarns in weaving (Interviewer 2, weaver, 2023). While using that they were seated in bare land. They are now seated on a bench or a chair. At that time, bamboo trees, plants, and tree pieces were used to make the heald shaft and reader (Interviewer 8, weaver, 2023). Tools and equipment are made by materials which are available in surrounding areas. In Eastern cultures, people are not seen as distinct from nature, and there is a traditional tie between nature and culture [33]. Yarns would frequently break while weaving intricate designs with those instruments, Therefore, they switched to a different machine that was created by altering the saree-making machine and required a larger heald shaft and reader. To wind threads they still use traditional machinery and new machinery. In the Ancient technological museum of Polonnaruwa, Sri Lanka, all information and sample machines were made according to their instructions. List of traditional tools are Hadawa, Alwale machine; Aluwa, Poruwa, and Koora patta.

- **Weaving patterns**

'iripath' design are made by the pattern 'pattini'.
'labaya' is made by the pattern 'depath'.
'kothmul' pattern is added at the four corners of the Cloth.
See the splendour and grandeur of the dumbara weaving!
(Poem 6: a poem describes Weaving patterns/English translation by author)

Traditional techniques of indigenous designs are recited in the poem six. This poem reveals original characteristics of Dumbara weaving. "Oral tradition focuses on recollections of the past that extend beyond one's lifetime [17], p. 19". A distinguished pattern weaves each design. The pattern resembling the goddess 'pattini' is difficult and hardly weaves these days. 'iripath' means slender lines that is the preliminary stage of developing the pattern 'pattini'. This is also called 'pannama'. 'depath' and 'kothmul' patterns are traditional.

The pattern 'Mudiyam' was made by looking at the sun.
The pattern 'Kanhida' was made by looking at the moon.
The pattern 'Kinihira' was made by looking at the strength of wind, The pattern 'Pasyale' was made by looking at the mountains.
(Poem 7: a poem describes the origin of patterns /English translation by author)

The poem seven describes the original inspiration sources for some traditional patterns. The sun, moon, wind, and mountains inspire patterns. The pattern 'mudiyam' is made by inspiring the sun. The pattern 'kanhida' is made by inspiring the moon. The pattern 'kinihira' is made by inspiring wind. The pattern 'pasyala' is made by inspiring mountains. Nine weaving patterns are described in poems: pattini, depath, thunpath, mudiyam, kanhida, kinihira, pasyale, depath and pannan.

- **Design, Motifs, and Colour**

Flowers, Animals, and geometric designs that are identical to the history and tradition of Dumbara weaving are described in poems.

Little flower petal, small flower, 'kathuru' flower.
Let the eight-petalled flower be in the middle of the petalled flower.
The plait, and knot work of plait, the 'domba' flower, the papaya flower.
Dumbara pattern has several flowers.
(Poem 8: a poem describes Flower types/English translation by author)

Poem eight describes designing a cloth with different flowers. Coomaraswamy stated [2], p. 235 that "eight-petalled flowers, little flower, and little flower petal, 'kathuru' flower and 'domba' flower are found on old cloths". Coomaraswamy [2], p. 94 observed 'karuthu' flower in early nineteenth century practice and explained that 'kathuru' flower is doubtless called from its crossing petals resembling scissors or crossed sticks. Papaya flower is a recently developed design. The eight-petalled flower is weaved in the centre of the petalled flower, as stated in the second line of the verse. Variety of plait works are considered as geometric designs. Coomaraswamy [2], p. 94 states that 'plait forms in early nineteenth century practice are very characteristic and logical including knot work/chequer work'.

The poem 5 also explains the composition of design elements that are handed down from generation to generation orally. This poem resembles early nineteenth century design compositing methods [2], p. 80. The poem shows that curtain traditional methods of combining flowers, animals, or geometric elements. The poem reveals that followers and animals need to be mixed in design composition. The first cross, flower of cross, knop and flower, flower called 'pana' are woven. Cross indicates by [2], p. 106 that 'how small equal armed crosses are derived from four petalled flowers have been indicated'. Knop and flower from a damascened lock, is modern but traditional [2], p. 102. Then, a line of wood pigeons and birds are woven between them. After that line-by-line Lion, lion with an elephant's head, elephant, tusker is woven. Interviewer 2 remembered that the weavers father used to weave beautiful clothes by adding birds such as pigeon, spotted doves. Now, birds are hardly woven. Big animals such as elephants, lion are placed in the centre.

Ten flower designs are described in poems: little flower petal, small flower, 'kathuru' flower, eight-petalled flower, petalled flower, 'domba' flower, the papaya flower, flower of cross, knop and flower, flower called 'pana'. In addition, Dumbara weaving displays six-petal flowers, ten-petal flowers, bunches of eight-petal flowers (Fig. 4), and bunches of six-petal flowers. Some of the traditional weavings

Fig. 4 Eight-petaled flower cluster (courtesy Mr. Priyantha Yapage, Kotuwelanda, Thalagune, Ududumbara)

mentioned in the verses are no longer visible. Weavers state that six-petal flowers, ten-petal flowers, eight-petal flowers, and six-petal flowers are more beautiful than the papaya flower. It is simple and easy to weave.

Parrots and hanging parrots are woven around the cloth.
The hill mynah is woven around the cloth.
lovable pattern called 'sara lapu' is weaved in the middle of cloth.
I weave a cloth to visit the king.

(Poem 9: a poem describes bird types and composition of design elements/English translation by author)

The poem nine explains the composition of design elements handed down orally from generation to generation. Parrots, hanging parrots are woven around the cloth. Lovable big petals are weaved in the middle of cloth. This cloth is weaved to visit the king. The cloth which was used by Kings of the Kandyan Kingdom (fourteenth century to 1815). Types of cloth which were used by the high-social class. The best quality that comes from tradition is for the King. The best quality has been the trend and luxury for ever since in Sri Lankan tradition. Indigenous cultural textiles show potentials towards sustainable development. It is seen that some native terminology such 'sara lapu' is forgotten and gone away from the practice.

Birds, emerald doves and hill mynah are woven.

Sparkling stars surround Sun and moon.
Peacocks in gold and pen are woven.
The fabric was woven using the kathuru' flower.
(Poem 10: a poem describes motifs on the fabric/English translation by author)

The poem Ten explains the motifs on the fabric include birds (Fig. 5), wood pigeons, hill mynah (uttering very loud musical whistling sounds), pen, and golden peacocks. Stars encircle the sun and moon, and the fabric is woven in the pattern of 'kathuru' flower (flower is called from its crossing petals resembling scissors or crossed sticks).

Each family possesses a unique aesthetic appreciation of colour mixing and patterning. Interviewer 1 (weaver, 2023) states that each family has identity in colour mixing design application. Interviewer 2 (weaver, 2023) said that they can identify the weaver by looking at wall hangers or cushions. Interviewer 7 (weaver, 2023) describes that they have created some modern designs derived from traditional designs.

Animal designs are described in poems such as elephant (Fig. 6), tusker, lion, wood pigeons, emerald doves, hill mynah, pen, golden peacocks, parrots, and hanging parrots. Due to the predominance of lions, elephant-lions, elephants, and tuskers in modern designs, birds are no longer visible.

Birds, pigeons, hill mynahs are weaved.
pen and golden peacocks weaved together.
A web of stars surrounded the sun and moon.
a cloth is weaved according to the patterns of depath.
(Poem 11: English translation by author)

This poem describes a weaving pattern called 'depath weave' and the designs and layout of the cloth. Birds, wood pigeons, hill mynahs, pen, and golden peacocks are

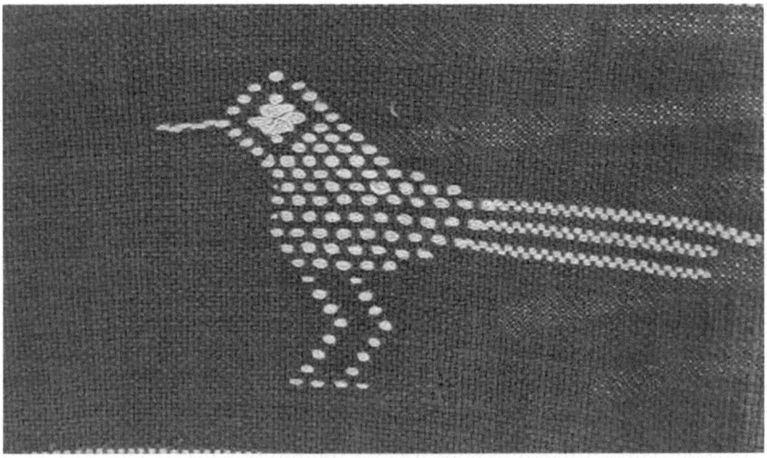

Fig. 5 Bird (courtesy Mr. Y. G. Premachandra, Kotuwelanda, Thalagune, Ududumbara)

Fig. 6 Elephant (courtesy Mr. Priyantha Yapage, Kotuwelanda, Thalagune)

mentioned. The sun, moon, and web of stars are also woven. Weavers are skilful in weaving traditional geometric shapes "gal-binduva" (set gem), knot work of plait. Six geometric shapes are described in poems: the plait, and knot work of plait, the sun, moon, and stars (Figs. 7 and 8).

- **Product Categories**

Fig. 7 The plait (courtesy Mr. Priyantha Yapage, Kotuwelanda, Thalagune, Ududumbara)

Fig. 8 Knot work of plait (courtesy Mr. Priyantha Yapage, Kotuwelanda, Thalagune, Ududumbara)

Products are produced as on-loom finished products, not as fabrics. In the past, bathing cloth called "diya-kacciya", bedspreads, pillowcases, cushion covers, sarees, and bags were made. Today, cushion covers, tablecloths, and wall hangings are famous. In the past, kings and nobles mostly used these products, and now are available in high-end handicraft shops in Colombo and the main cities of Sri Lanka. Textiles are made by artisans working in their rural homes. Due to the uniqueness of this tradition tourists are interested in it and the village becomes their famous tourist destination. Today weavers use business cards, technology to develop an overseas customer base. Their ancestors' creations are now exhibited in Sri Lanka and United Kingdom museums. Master weavers in this heritage village of Thalagune have been awarded the 'Kalabhooshana' State Award for their excellence in craft.

The pride of a weaver is displayed by weaving the oldest cloth type called "diya-kacciya" or bathing cloth worn by the Knadyan royalty and nobles. The Thalagune artisans claim that loin cloth/bathing cloth "diya-kacciya" is their precious cloth made only by their ancestors. This dress is a significant dress during the Kandyan Kingdom of Sri Lanka (fifteenth century to 1815). Interviewer 1 (weaver, 2023) states that "A sample of "diya-kacciya" that was woven by my great grandfather s at the display of the Museum of India". Interviewer 2 (weaver, 2023) states that "The most precious cloth of all that we have been able to weave is called "diya-kacciya". That product's weaving is quite complex, and its technology was incredible. That garment was woven in one piece and had no seams". Interviewer 4 (weaver, 2023) states that "A "diya-kacciya" is at the display of the Museum of Kandy, Sri Lanka and a museum of England was woven by my great grandfather whose information is also displayed". Interviewer 6 (weaver, 2023) states that "The king was pleased with my great grandfather's weaving skills then got woven some clothes for the King".

> Lotus to the side of the head and woven pillow.
> Lotus to the side of the leg and woven pillow.
> Woven a cloth to cover the bed.
> The pattern 'thunpath' is used to complete the cloth.
> (Poem 12: English translation by author)

By applying lotus to the side of the head and leg, the weaver means that they were weaving pillows with respect (interviewer 2 weaver 2023). Respect and care affect well-being [6]. The last two lines of the verse describe the weaver making a bedspread using the Thunpath pattern. Interviewers state that the traditional product category has been changed today. It is seen that they have increased their product category to suit current customer demand. They produce bags (Fig. 9), cushion covers. However, it is significant that they still produced traditional historic itinerary such as "diya-kacciya" (loincloth/bathing cloth) to showcase their pride and inherited skills. Due to numerous developments in weaving and the commercialisation of society, artisans produce designs in response to market demand. Interviewer 6 (weaver, 2023) states, "Price depends on design. When a consumer requests a low-cost product, we create it with an easily weavable pattern. Some buyers are not concerned with price. They pay for any aesthetically pleasing design." (Interviewer 6, weaver, 2023). Traditional artists have understood the concurrent demand of the customers. Weavers describe contemporary use of flower types as below. Interviewer 2 (weaver, 2023) states, "These days, eight-petaled, ten-petaled, and six-petaled flowers are used. Those are three beautiful flowers. These furthermore functioned as a bouquet of eight-petaled flowers. Add four or five flowers. Another one is known as a cluster of six-petal flowers" (Interviewer 2, weaver, 2023). Tradition changes, and artists adapt with happiness [42].

The Thalagune community upholds and spreads cultural values through their tangible culture and heritage. It has long been understood that landscape and memory are related. From their parents or grandparents, all the artisans learnt how to weave. Inherited Dumbara rata weaving was passed down orally from generation to generation, preserving its unique qualities. They recalled received information. They followed the same in the past while taking pleasure in their profession daily. At that time this community used lyrics to guide weavers through their work and serve as a reminder of the proper practices. They continue to navigate modern life with their hereditary wisdom. The methods, tools, weaving techniques, design and motifs used in the verse are specific to this community, aiding in forming their identity. They continue to weave traditional Dumbara patterns as it is, and by drawing inspiration from modern designs, they also produce contemporary designs. Today, they create designs in response to market demand.

Fig. 9 Bag with plait and knot work in the middle

5 Conclusion

The ability to make a livelihood from one's skill can be considered to reduce inequality in the textile industry. The identity of traditional weaving positively minimises the disparity in the textile industry. A new conventional interpretation of quality substitutes the meaning and application of luxury. The quality that comes from oral tradition is the trend and luxury. Indigenous cultural textiles show potential towards sustainable development. Luxury brands look for identity and uniqueness. The process of design offers the potential to provide firm alternatives. Traditional weaving shows unique aesthetic expression as they have been inherited from generation to generation. It suggests developing local perspectives, using traditional artisans' skills, and letting them earn. When designers join the instinctive with the mechanical, the hand-crafted and the technological, the result can be magical. Direct and indirect involvement of traditional weaving artists in the industry is expected.

Today Dumbara indigenous weaving has become both cultural symbols and commercial assets. The commercialisation of traditional crafts is a part of the survival mechanism, as well as of the development of artisans and communities. Weavers have improved their quality of life by understanding and realising their potential in the art and craft of traditional hand-loom weaving. Their survival pattern and characteristics are found. The way of life contains the capacity for continuance, which is bound to the inheritance memory. Folk poems, stories, ritual recites, and beliefs show their valuing system of life and occupation. The research findings help to understand the relevance of traditional arts in contemporary social life. Reinforces knowledge aimed to support the introduction of crafts and skills to stakeholders. Well-being is correlated with harmony, sharing, caring, peace, and empathy of the human capital collectively. Qualities which protect the community need to be carefully facilitated.

Culture, tradition, and sustainable rural life create a traditional art and craft identity. Their culture offers a common destiny based on shared practices, memories, and values. In traditional civilisation, survival and sustainability were intertwined. The sustainability of the art and craft of a traditional community depends on the maintenance of resources. Learning methods in traditional art and craft are repetition, prolonged engagement, and collaboration with artists. The value of re-discovery and promoting traditional crafts in their oldest perceivable form helps to maintain their well-being and knowledge.

Care holders are elders of indigenous traditions who hold the original art. They protect the basic indigenous knowledge. Elders are active resource persons who eagerly hold the responsibility of teaching to the next generation. The novice learns to repeat traditional patterns, designs and motifs, colours and gradually develops innovation from fundamental knowledge. Oral tradition is the resource pool, and it serves as a better learning tool which bridges past and present. Nine weaving patterns are described in poems: pattini, depath, thunpath, mudiyam, kanhida, kinihira, pasyale, depath and pannan; association with traditional colour, design and motifs, product categories are recorded. There is a correlation between the traditional craft's well-being and the weaver's prolonged engagement. It is observed that when practice is decreased, the traditional knowledge is forgotten. Traditional knowledge, craftsmen and their living village are the main actors to be active in future development plans in textile industry.

References

1. Burt EL, Atkinson J (2012) The relationship between quilting and well-being. J Public Health 34(1):54–59. https://doi.org/10.1093/pubmed/fdr041
2. Coomaraswamy A (1956) Meadival Sinhalese art. National Museum, Sri Lanka
3. Croll E (2006) China's new consumers: social development and domestic demand. Routledge
4. Daskon C, Binns T (2010) Culture, tradition and sustainable rural livelihoods: exploring the culture-development interface in Kandy, Sri Lanka. Commun Dev J 45(4):494–517. https://doi.org/10.1093/cdj/bsp019
5. de Silva KM (2005) A history of Sri Lanka, The first Sri Lankan edn. Vijith Yapa Publications, Colombo
6. Dias RM, Ogle JP, Diddi S (2020) Constructing cultural identity through weaving among Ri-Bhoi women weavers: a symbolic interactionist approach. Fash Text 7:1–21. https://doi.org/10.1186/s40691-020-00220-x
7. Fave AD, Kocjan GZ (2016) Well-being in the arts and crafts sector. In: The Wiley Blackwell handbook of the psychology of positivity and strengths-based approaches at work, 508–526. https://doi.org/10.1002/9781118977620.ch26
8. Fong GS (2004) Female hands: embroidery as a knowledge field in women's everyday life in late imperial and early Republican China. Late Imperial China 25(1):1–58. https://doi.org/10.1353/late.2004.0007
9. Giovannetti G, Velucchi M (2022) Gender discrimination and firm survival: a multilevel approach for EU textile companies. SN Bus Econ 2(9):142. https://doi.org/10.1007/s43546-022-00315-1

10. Gunti Amaravathi DBB (2023) Socio-economic position of hand-loom households in India. J Surv Fish Sci: 688–701. https://doi.org/10.17762/sfs.v10i1.650
11. Hallisey C, Hansen A (1996) Narrative, sub-ethics, and the moral life: some evidence from Theravāda Buddhism. J Religious Ethics: 305–327. https://www.jstor.org/stable/40015212
12. Hill FW (2004) Passing on traditional knowledge. In: Indigenous knowledge conference proceedings. Pennsylvania State University, University Park, PA
13. Inglehart R, Baker WE (2000) Modernisation, cultural change, and the persistence of traditional values. Am Sociol Rev: 19–51. https://doi.org/10.2307/2657288
14. Kenning G (2015) "Fiddling with threads": craft-based textile activities and positive well-being. Textile 13(1):50–65. https://doi.org/10.2752/175183515x14235680035304
15. Kuhlman T, Farrington J (2010) What is sustainability? Sustainability 2(11):3436–3448. https://doi.org/10.3390/su2113436
16. Lambert M (2014) The lowest cost at any price: the impact of fast fashion on the global fashion industry
17. Li N (2020) History, memory, and identity: oral history in China. Oral Hist Rev 47(1):26–51
18. Mayer H, Knox PL (2009) Pace of life and quality of life: the slow city charter. In: Community quality-of-life indicators: best cases III. Springer Netherlands, Dordrecht, pp 21–40. https://doi.org/10.1007/978-90-481-2257-8_2
19. McFarlane TR (2007) The contribution of taewa (Maori potato) production to Maori sustainable development (Doctoral dissertation, Lincoln University)
20. Memon H, Ranathunga GM, Karunaratne VM, Wijayapala S, Niles N (2022) Sustainable textiles in the past "wisdom of the past: inherited weaving techniques are the pillars of sustainability in the hand-loom textile sector of Sri Lanka." Sustainability 14(15):9439. https://doi.org/10.3390/su14159439
21. Mignamissi D, Nguekeng B (2022) Trade openness-industrialisation nexus revisited in Africa. Econ Chang Restruct 55(4):2547–2575. https://doi.org/10.1007/s10644-022-09401-8
22. Minton EA, Tan SJ, Tambyah SK, Liu RL (2022) Drivers of sustainability and consumer well-being: an ethically-based examination of religious and cultural values. J Bus Ethics 175:167–190. https://doi.org/10.1007/s10551-020-04674-3
23. Nurse K (2006) Culture as the fourth pillar of sustainable development. Small Stat Econ Rev Basic Stat 11:28–40
24. Ouyang X, Jian Q, Jiang Z (2022) Energy efficiency of the industrial sectors in Beijing-Tianjin-Hebei urban agglomeration: does technological gap matter? Environ Sci Pollut Res 29(34):52263–52276. https://doi.org/10.1007/s11356-021-17522-2
25. Pressman SD, Kraft T, Bowlin S (2020) Well-being: physical, psychological, and social. In: Encyclopaedia of behavioral medicine. Springer International Publishing, Cham, pp 2334–2339. https://doi.org/10.1007/978-3-030-39903-0_75
26. Ranathunga GM (2018) Art and tradition of Sri Lanka-volume 01: music of Sri Lanka
27. Riley J (2008) Weaving an enhanced sense of self and a collective sense of self through creative textile-making. J Occup Sci 15(2):63–73. https://doi.org/10.1080/14427591.2008.9686611
28. Roberts RE (2020) Qualitative interview questions: guidance for novice researchers. Qual Rep 25(9)
29. Senanayake SGJN (2006) Indigenous knowledge as a key to sustainable development. J Agric Sci 2(1)
30. Sethi B (2023) Cultural sustainability: influence of traditional craft on contemporary craft cross-culturally
31. Shils E (1981) Tradition. University of Chicago Press
32. Srivastava J (2023) The problems and challenges of the hand-loom co-operative societies in Varanasi district. IJFMR Int J Multidisc Res 5(2)
33. Taylor K (2008) Landscape and memory: cultural landscapes, intangible values and some thoughts on Asia
34. Utley FL (1961) Folk literature: an operational definition. J Am Folklore 74(293):193–206. https://doi.org/10.2307/537632

35. Wickramaarachchi N, Grodach C, Ranathunga GM, Ratnayake R, Karunarathne PVM (2022) Craft industries and ribbon development: place change along the Colombo-Kandy Road, Sri Lank. Int Plann Stud 27(2):196–212. https://doi.org/10.1080/13563475.2022.2038545
36. Wheatley D, Bickerton C (2017) Subjective well-being and engagement in arts, culture and sport. J Cult Econ 41:23–45. https://doi.org/10.1007/s10824-016-9270-0
37. Wills E (2009) Spirituality and subjective well-being: evidences for a new domain in the personal well-being index. J Happiness Stud 10:49–69. https://doi.org/10.1007/s10902-007-9061-6
38. Wijayapala UGS, Alwis AAP, Ranathunga GM, Karunaratne PVM (2022) Evolution of Sri Lankan textile education from ancient times to the 21st century. In: Textile and fashion education internationalization: a promising discipline from South Asia. Springer Nature Singapore, Singapore, pp 119–144. https://doi.org/10.1007/978-981-16-8854-6_7
39. Wosk J (2017) Manus × machina: fashion in an age of technology by metropolitan museum of art. Technol Cult 58(1):257–263. https://doi.org/10.1353/tech.2017.0013
40. Yanpisit K, Surangkul S (2022) Phrae-wa studies and driving towards sustainable development goals by university for local development. Asian J Trad Innov Arts Text 1(1):55–72
41. Yeh JHY, Lin SC, Lai SC, Huang YH, Yi-Fong C, Lee YT, Berkes F (2021) Taiwanese indigenous cultural heritage and revitalisation: community practices and local development. Sustainability 13(4):1799. https://doi.org/10.3390/su13041799
42. Zhang X, Li Y, Lin J, Ye Y (2021) The construction of placeness in traditional handicraft heritage sites: a case study of Suzhou embroidery. Sustainability 13(16):9176. https://doi.org/10.3390/su13169176
43. Zuallcobley RW, Nik MZA (2020) Safeguarding indigenous knowledge and cultural tradition while promoting sustainable tourism. Asian J Arts Cult Tour 1(2):22–25

Heritage of Living Entity and Living Skills: Advancing Sustainability Goals 1 and 4 in Handloom Textiles

P. V. M. Karunaratne, U. I. Jayalath, and K. P. N. Bhagya

Abstract Handloom textiles in the Southern Province are a testament to the cultural legacy and traditional knowledge passed down through generations. With the continued practice of traditional design methods and techniques, these textiles stand as a 'Living Heritage.' This term captures the intangible cultural heritage of handloom textiles, encompassing traditions, knowledge, and skills inherited over time. The Textile Department of the Southern Province consistently motivates and encourages the skill development of handloom textile weavers. As a result, these weavers are constantly engaged with 'Living Skills,' showing a readiness to learn at all times. Recognizing the industry's interconnectedness with societal facets, such as culture, environment, and economic growth, a holistic approach to promoting and achieving sustainability in the handloom textile sector. The Textile Department of the Southern Province adheres to the Sustainable Goals by following a Strategic Plan. This Strategic plan seeks to balance the requirements of the various elements of Sustainable Goals. The aim of the study is to identify how handloom textile weaving heritage prevails as a living entity and living skills in the Southern Province of Sri Lanka toward advancing Sustainability Goals 01 and 04. Despite the challenges posed by the current economic situation and administrative shortcomings, the Department and its management recognize and value the weavers and the broader workforce for their significant economic contributions. These weavers and textile designers are valued as the vital human capital of the industry. This chapter emphasizes the alignment with SDG 01, which focuses on enhancing people's skills for improved livelihoods, and SDG 04, centered on Learning and Education. It underscores the importance of collaborating with Universities to leverage their educational resources and training for the weavers. This partnership also facilitates the dissemination of information to the broader local community. This research employed qualitative research

P. V. M. Karunaratne (✉) · U. I. Jayalath · K. P. N. Bhagya
Department of Textile and Apparel Engineering, University of Moratuwa, Moratuwa, Sri Lanka
e-mail: virajinik@uom.lk

U. I. Jayalath
e-mail: umajayalath111@gmail.com

K. P. N. Bhagya
e-mail: nilminibhagya94@gmail.com

G. M. Ranathunga (ed.), *Textiles of Sri Lanka*, SDGs and Textiles,
https://doi.org/10.1007/978-981-97-0569-6_3

methods. The Primary data was collected through participant observation and semi-structured interviews. The researchers directly observed Departmental documents to glean further information. Administrative Officers, Weavers, Coordinating Officers of the Department, and Instructors were interviewed to gain insights into the experiences, perspectives, and viewpoints of SGD 01, which is vital to nurturing people's skills to attain a more prosperous and enriched life. SGD 04 focused on learning and education in the textiles industry, which has held a pivotal role in societies over the ages, offering a wealth of opportunities for exploration. Universities focusing on textiles can significantly impact the sector by offering free educational tools, training, and sharing expertise within the local community. Policymakers, academics, and government authorities will benefit from this study.

Keywords Handloom Weaving · Living Heritage · Living Skills · Sustainability Goals · Weavers

1 Introduction

In September 2015, member countries of the United Nations (UN) came to a consensus on the Seventeen Sustainable Development Goals (SDGs) as part of their commitment to the 2030 Agenda for Sustainable Development [119]. The SDGs represent a transformative approach to global sustainability and development. They differ notably from previous development paradigms in terms of their intended universality, expanded thematic range, introduction of key players, and emphasis on financing for development. Despite skepticism, the 2030 Agenda has garnered substantial traction, with many countries globally adopting the SDGs [61, 102]. However, the universality of the SDGs presents challenges. While inclusivity in crafting and executing these goals is commendable, it sometimes limits flexibility—essential for adjusting the SDGs to specific local situations. This challenge resonates with broader post-colonial concerns about global impositions. It underscores a vital tenet: for development to be truly effective, it must be conceptualized locally, resonate with local visions, and not be perceived as an external mandate.

While the SDGs are not without shortcomings, they offer communities a holistic framework, fostering critical and inventive thinking as we contemplate improved futures. An OECD analysis of green recovery measures indicates that governments understand the value of "job creation and economic boost via ecosystem rejuvenation." Yet, there's a noticeable lack of focus on skill development. The OECD underscores that investments in education and training are pivotal for a seamless shift to a green economy that serves every community. In line with the UN's objective to "ensure no one is left out," everyone should have the chance to acquire the essential skills to move into and prosper in more environmentally friendly industries [7]. Technical know-how, proficiency, and talents known as "green skills" enable the efficient application of green technology and procedures in the workplace. They draw from various knowledge, values, and attitudes to facilitate ecologically sustainable decision-making at work and in life [120].

Crafts, as living traditions, are enriched with traditional skills that evolve. While rooted in heritage, these skills are often adapted and transmitted in response to societal, cultural, or economic shifts, showcasing their dynamic and ever-changing nature. They are weaving embeds in their 'habitus.' This term, introduced by a sociologist [12], denotes the shared principles inherent within a specific structure or environment type. According to Johnson [59] these guidelines "produce and structure actions and perceptions that can be tangibly aligned with their results. "These concepts also have "permanent dispositions," and "regulated improvisations." They are "durably installed." Furthermore, in anthropology, apprenticeship has been "re-identified as a key area linking theoretical knowledge with practical application" [71]. Such studies have significantly benefited from the "situated learning" paradigm developed by Lave [65].

2 The Rationale of the Chapter

Living heritage is a continuing inspiration for the weavers' cultural ethos. They produce handwoven fabrics as an artisanal object of Sri Lankan heritage and Culture. Innovative weavers exquisitely combine regional patterns and techniques. They have woven symbols and motifs in the textile to foster a sense of identity among the local community. Weavers support communal harmony and integrity, which is essential to sustainability. The handloom industry supports local economies, creates jobs locally, and encourages participation in the country's cultural heritage.

This chapter delves into the connection between the sustained practice of heritage and traditional skills in the handloom textile sector. This exploration is in light of increasing academic and societal debates suggesting that intangible cultural heritage supports the Sustainable Development Goals, especially by enhancing the genuine nature of goods and communities. Proponents of this position argue that certain commonalities of Sustainability Goals create such capacity development and traditional skills forwarding to the fore. I hone in on one facet of this argument: intangible cultural products frequently involve tradition and crafts skills and give cultural value for development and vice versa to the producer.

Nevertheless, this line of reasoning lacks solid empirical evidence and a deep analysis of the intricacies of creating social and cultural value.

3 Interpretation of Heritage: Living Heritage

Any component of a society's past or present values and wishes to preserve them for future generations is considered part of its cultural heritage. It is frequently described as things, locations, or practices that have a long history, are rooted in tradition, or have historical significance [50]. In its broadest meaning, UNESCO defines cultural heritage as the legacy of tangible items and intangible characteristics of a group or civilization that have been passed down from earlier generations, conserved in the present, and given to the benefit of subsequent generations [81]. It is not always about things from the past. Even if they are brand-new or have undergone recent alteration, certain things, places, and practices may still be culturally significant to the current generation and have the potential to be part of our cultural heritage [90]. Cultural heritage refers to valuable objects or events deemed necessary by individuals or organizations. Viewing heritage as a pluralistic concept is insignificant [62]. Cultural heritage gives each unique place distinctive characteristics and is a priceless source of human experience [114]. According to the UN (1995), every aspect of the heritage of indigenous populations has its designated custodians. These caretakers can be the entire community, a particular family or clan, a group or organization, or individuals specifically trained or chosen to oversee it [32]. Heritage reflects inherited and current concerns about the past [50]. Heritage is the knowledge that forms economic and cultural capital, crucial in an age of multiscale space analysis [44]. According to scholars, heritage can be defined as tangible, immovable resources (buildings, rivers, and natural regions), tangible, movable resources (items in museums, papers in archives), or intangibles like values, conventions, rites, lifestyles, and experiences like festivals, the arts, and cultural events [113, 127]. Smith Smith [105] emphasizes heritage as a dynamic, evolving concept requiring community participation, identity building, and social cohesion for its development.

Heritage interpretation effectively manages, learns about, and communicates historical sites [86]. Uzzell [121] identified that understanding heritage is vital for strengthening visitors' sense of place and identity. If this happens, the interpretation must be designed and created with this outcome in mind. Heritage marketing has often been viewed as a skeptical endeavor to tap into and satisfy the public's desire for comforting and nostalgic representations of history. Nevertheless, it's undeniable that interpreting heritage can enhance one's understanding, insight, and esteem for specific eras and locations. When approached with this intent, interpretation can indeed be beneficial [121]. Harvey (2015) views legacy as a dynamic process, evolving as people and societies reinterpret and shape it. Meskell (2013) highlights UNESCO's neglect of the dynamic nature of heritage, arguing that it evolves into a concept renegotiated by communities and stakeholders. They call for an inclusive, participatory approach to heritage management and preservation, viewing it as a resource for community empowerment and sustainable development (Meskell 2013). These studies emphasize the significance of understanding legacy as a living thing continuously changing and molded by people and communities.

4 The Role of Heritage

The role of heritage lies in the essential act of maintaining and conserving. It primarily focuses on doing all possible to delay the natural laws of deterioration on tangible artifacts to ensure the transmission of its vital heritage messages and values to future generations [80]. Heritage assets and associated values anchor communities to look to the past, comprehend the present, and plan for the future [107]. Harrison [50] says that heritage practices are customs and habits that, while intangible, inform who we are as a collective and help create our collective social memory. He demonstrates that engaging with heritage items (like artifacts, structures, locations, and sceneries) and heritage activities (such as languages, music, community celebrations, and the conservation of past objects or memories) influences our perceptions of our history, current times, and what lies ahead.

5 People Modify Heritage Through Active Participation

This paragraph explores how people actively participate in and modify their ancestry and Culture, recognizing that heritage involves engaging with the present and future and advocating for creative transformation. Heritage is more than conserving the past; it also entails engaging with the present and the future. They support altering one's heritage through a creative transformation process in which one actively alters and creates one's past [126]. Cultural legacy is a product of both the past and the present; according to Silverman [104], historical events and contemporary influences shape cultural heritage. "They advocate for people to adapt their cultural heritage to meet their current needs and to reflect their identity and beliefs. Moore-Cherry and Whelan [75] emphasize adapting cultural legacy for everyday needs and identity. Lam's [64] study examines Lijiang residents' adaptation of cultural heritage for tourism, resulting in economic benefits but loss of authenticity and displacement. Addressing potential negative implications is crucial [132].

Additionally, Endresen [40], Del Barrio [35], and Mortara [76] highlighted that there are several reasons for the public and private sectors to be involved in the management of cultural heritage. UNESCO [115] highlighted the significance of community involvement in heritage and Culture. UNESCO stated that active community participation in heritage conservation and adaptation requires sustainable cultural heritage management. Sustainable cultural heritage management includes protecting the physical heritage sites and involving nearby communities in decision-making. Smith [105] and Waterton [126] investigate how people portray their cultural history and change it to meet their modern requirements. They suggest cultural heritage representation is a negotiation process between stakeholders, such as local communities, heritage specialists, and visitors. Tong Lam [64] analyzes how residents in Lijiang, China, adapt their cultural heritage to meet tourism needs. While this has benefited the region economically, it has also resulted in the loss of some cultural authenticity and the displacement of local communities. Finally, the research argues that people change their heritage/culture through creative transformation, community participation, negotiation, and representation. While this can result in positive improvements, addressing the potential negative implications of changing cultural heritage is also necessary. The statement also emphasizes the importance of heritage resources in ensuring community continuity, rejuvenation, and advancement. Additionally, improving the quality of life through cultural heritage management is highlighted, as is the significance of these resources in nurturing and maintaining individual and collective identities.

6 The Link Between Heritage and Society: Past and Today

Cameron [25] emphasizes understanding heritage's temporal and spatial boundaries, which may fade or change due to future social and cultural conditions. He states that cultural heritage can be an economic advantage, a tourist attraction, a source of identity, and a component of social cohesion. In 2018, the European Year of Cultural Heritage (EYCH) saw EU Member States organizing events and conferences and allocating funds for cultural heritage preservation and development–EYCH positively impacts economic, social, and environmental consequences [69]. The link between heritage and society encompasses heritage conservation,it is maintaining cultural heritage, primarily focusing on doing all possible to delay the natural laws of deterioration on tangible artifacts to ensure the transmission of its vital heritage messages and values to future generations [69].

In 2018, Pasikowska-Schnass highlighted that cultural heritage is not just about costs or solely tied to identity-related matters. Instead, it offers numerous benefits across three sectors, aids in forming social capital and a communal sense of identity, and fosters societal unity.

- Economic—Cultural heritage can serve as a model for economic production, positively influencing job creation and even extending its impact to other sectors.
- Social—Cultural heritage fosters unity, inclusivity, cohesiveness, and community participation.
- Environmental—Cultural heritage supports the sustainable evolution of landscapes. Cultural legacy is essential in society because it supports cultural resilience—specifically, how it has adapted and flourished in the past [53].

7 The Present Status of Prevailing Heritage in Any Society

Heritage is a valuable cultural resource, generating billions of euros in revenue and millions of jobs annually [80]. Following the ratification of the 1972 Convention, UNESCO established the World Heritage List, which features exceptional heritage sites that require safeguarding [69]. Heritage and Culture influence the daily lives of residents of every town. Additionally, heritage is considered one of the most accessible resources for tourism [113]. Cultural legacy may be a source of wealth and a source of identity, and it can also promote social cohesion [81]. However, it is becoming more apparent that cultural heritage's value depends on its contributions to society.

Furthermore, heritage stands out as one of the prime attractions for tourists [113]. Cultural heritage's value depends on its contributions to society. Recognizing and embracing changes in cultural heritage can improve resilience and risk preparedness. Preservation should inspire uncertainty and human evolution, enabling adaptation to changing circumstances [53]. Cultural heritage fosters citizenship, belonging, and identity [83].

Holtorf [53] raised a question and asked how people can learn to recognize and embrace the changes that have emerged in the transition of cultural heritage over time to improve their cultural resilience and risk preparedness. People who are more resilient and able to handle change, including the loss of particular cultural manifestations, can better cope with hardship in their own lives and adjust to new situations. With this in mind, perhaps cultural heritage preservation should not emphasize the inherent worth and timeless features of the artifacts of the past that left behind us. Instead, by serving as inspiration for the need to embrace uncertainty and the capacity of humans to continue evolving, this cultural heritage may help us better adapt the legacies of the past to changing circumstances both now and in the future [53].

Present-day conceptions of citizenship, inclusion, and self-identity are formed and maintained based on culture and cultural heritage [89]. According to Di Pietro et al. [83], the formation of identities and habits in both individuals and groups is fundamentally influenced by culture. According to the Faro Convention of the Council of Europe, the value and potential of cultural heritage, if properly maintained, is an essential element of community development and a factor in greater quality of life in constantly evolving communities [88].

8 The New Movements in Safeguarding Heritage

Since the creation and adoption of the UNESCO Convention [117] on the Safeguarding of Intangible Cultural Heritage, there has been a considerable increase in the level of discussion on the nature and value of intangible heritage and the relevance and character of the legacy. The Convention evaluates societal issues and Indigenous cultural ambitions [106].

The process of preserving cultural legacy is known as heritage conservation, and its primary goal is to delay the natural laws of deterioration on physical items to preserve the transmission of its essential heritage messages and values to future generations [80]. Current heritage conservation practice activities, mainly carried out by conservation practitioners in museums, conservation laboratories, and monuments worldwide, widely involve implementing preventive, remedial, and restoration activities [80]. With 47 European member countries, the Council of Europe (CoE) concentrates its activities on human rights, the rule of law, and democracy, with the latter embracing Culture and cultural heritage as aspects of democratic administration. Its 1992 Valetta Convention for the Protection of Archaeological Heritage adopts the goal of protecting archeological heritage from illegal excavations and significant construction projects, and its 1985 Granada Convention for the Protection of European Architectural Heritage established the principles of European cooperation and policy coordination in architectural conservation [81].

9 Sustaining Heritage Through the Interpretation of Craft Textiles (Handwoven Objects)

Craft textiles, especially items made by hand weaving, are a significant element of cultural history worldwide. Maintaining this history is a complex endeavor that calls for striking a balance between maintaining traditional methods and designs and adjusting to shifting consumer preferences and economic realities. Through interpretation, artisan textiles can help preserve cultural history. Standard items give new interpretations and stories relating to current problems and concerns. Doing this makes the things more relevant to contemporary audiences, maintaining their worth and relevance. For instance, a study of Portuguese textiles highlighted their cultural and social relevance, utilizing them to explore topics like identity and globalization [72]. Kim et al. [63] connected traditional Korean textiles to modern environmental and ethical production challenges by interpreting them through the lens of sustainability [63, 72]. Collaboration between stakeholders is necessary to sustain history through artisan textiles and interpretation. The partnership covers collaborations between designers, makers, government institutions, regions, and nations. For instance, in a study by Santos et al. [96] Brazilian craftspeople and designers worked together to create new goods that fused traditional methods with cutting-edge design. Finally, maintaining a legacy through craft textiles requires careful consideration of social, economic, and environmental sustainability. Agrawal [2] studied handwoven textiles from India. They identified sustainability characteristics of their textile industry, such as using natural materials and dyes and supporting local communities. Maintaining tradition through interpreting artisan textiles necessitates a comprehensive strategy considering sustainability, collaboration, and interpretation. Thus, classic items can continue to exist and be helpful in a constantly evolving world.

10 SGD for Craft Sector in the World

The world of craft practices is one of the five domains included by the UNESCO Convention for the Safeguarding of the Intangible Cultural Heritage (ICH), established in 2003. Although the need to protect and sustainably manage ICH has consistently been recognized as critical, it gained global prominence after the Convention's creation and ratification by over 179 nations, including Sri Lanka [100]. The 193 UN member states' governments defined the 2030 Agenda for Sustainable Development in September 2015 [119]. Its central component is the 17 Sustainable Development Goals (SDGs), which comprise the most significant portion of the 2030 Agenda's 169 targets [119], encompassing the three pillars of sustainable development. They are a replacement for the Millennium Development Goals (MDGs), endorsed by the United Nations in 2000 and whose deadline was December 2015 [118]. They represent a new endeavor or commitment involving all nations in economic, social, and environmental sustainability for 15 years until 2030. In keeping with the spirit of our 2003 Convention, UNESCO wished to contribute to the discussion and make an effort to provide solutions to preserve and safeguard traditional textiles [116].

The importance of protecting Intangible Cultural Heritage (ICH) has been globally recognized since the initiation of the 2003 Convention. As outlined in its first article, one of the core elements of this Convention is safeguarding ICHs and establishing a List of ICHs in Need of Urgent Safeguarding. The Office of the Development Commissioner for Handicrafts and Handlooms under the Ministry of Textiles oversees this sector in India. This office is deeply aware of the industry's challenges and has consistently implemented policies and programs to ensure its sustainability, keeping with the goals of preserving intangible cultural heritage [101]. The handicraft industry can grow through a social process in which many people can participate as a source of income, a way to strengthen their abilities as entrepreneurs, and a way to promote the quality of their work [70, 78].

11 Initiatives for SGD in Sri Lanka

Sri Lanka was among the early signatories to the United Nations' Sustainable Development Goals (SDGs), encompassing 17 objectives such as SDG 1—No Poverty: SDG 2—Zero Hunger: SDG 3: Good Health and well-being: SDG 4—Learning and Education: SDG 5—Gender Equality: SDG 6—Clean Water and Sanitation: SDG 7—Affordable and Clean Energy: SDG 8—Decent Work: SDG 9—Industry and innovation: SDG 10—Reduced Inequalities: SDG 11—Sustainable Cities: SDG 12—Consumption and Production: SDG 13—Climate Action: SDG 14—Life Below Water: SDG 15—Life on Land: SDG 16—Peace, Justice, and Strong Institutions: SDG 17—Partnerships for the Goals:

The United Nations introduced the Sustainable Development Goals (SDGs) in 2015; since then, they have garnered popularity worldwide. To attain a more sustainable future, several nations, including Sri Lanka, have matched their national policies and activities with these objectives. Since endorsing, Sri Lanka has taken some actions to achieve the 2030 Agenda for Sustainable Development at home. These actions include creating a ministry solely dedicated to sustainable development to monitor and report on SDG progress, convening a select committee of parliament to provide political direction for SDG execution, and passing the Sustainable Development Act No. 19 of 2017 to enshrine the SDGs into law formally [20, 123]. The Sustainable Development Act identified six parameters in line with Vision [123] propaganda.

1. **National Integration and Vision 2025:** Sri Lanka aimed to integrate SDGs into its National Policy Framework to ensure a sustainable and robust economy by 2025 [123].

2. **Institutional Mechanism:** The establishment of the Sustainable Development Council (SDC) by Sri Lanka demonstrated its commitment to overseeing and driving the SDGs at the national level [109].
3. **Education and Skills Development:** The industry's commitment to SDG 4 (Quality Learning and Education) was evident through various initiatives such as the workshops and training programs
4. **Awareness and Participation:** Sri Lanka has organized various conferences and events, like the "Sri Lanka NEXT—A Blue Green Era" conference, to raise awareness and bolster SDG-related actions.
5. **Voluntary National Review (VNR):** Sri Lanka presented its Voluntary National Review at the UN in 2018, showcasing its SDG progress. The VNR can be valuable for detailed insights [20].
6. **Partnerships:** The country has fostered regional, national, and global partnerships to ensure the holistic realization of the SDGs.

12 The Study Context, Research, Methodology

The survey for this study was conducted in the three districts of the Southern Province of Sri Lanka in 2023. These three districts are Galle, Mathara, and Hambanthota, well-known centers of handloom weaving own their weaving and dye centers (Fig. 1). Each center employs one weaving instructor, but many are over 60 and rarely report to the centers. The number of weavers at each center is limited. A case study used interviews, observations, and document analysis to understand handwoven weaving in the Southern Province. This research delves into the growing prominence of the handwoven industry in the Southern Province as a notable destination for handloom weaving heritage. It emphasizes managing features that fortify Sustainable Development Goals, viewing the industry as a valuable economic asset. Locally, the textile sector thrives, ensuring the ongoing skill development of weavers who embody the cultural, sociological, and economic facets of weaving and its community impact. The investigation centers on communities or regions renowned for handloom weaving, considering factors like the number of weavers, traditional methods, historical relevance, and data accessibility. Table 1 illustrates the SWOT analysis conducted in the pilot study, which shows a tool to gauge an industry's position in the competitive landscape and guide strategic decision-making. It examines both internal and external elements, considering present conditions and prospects.

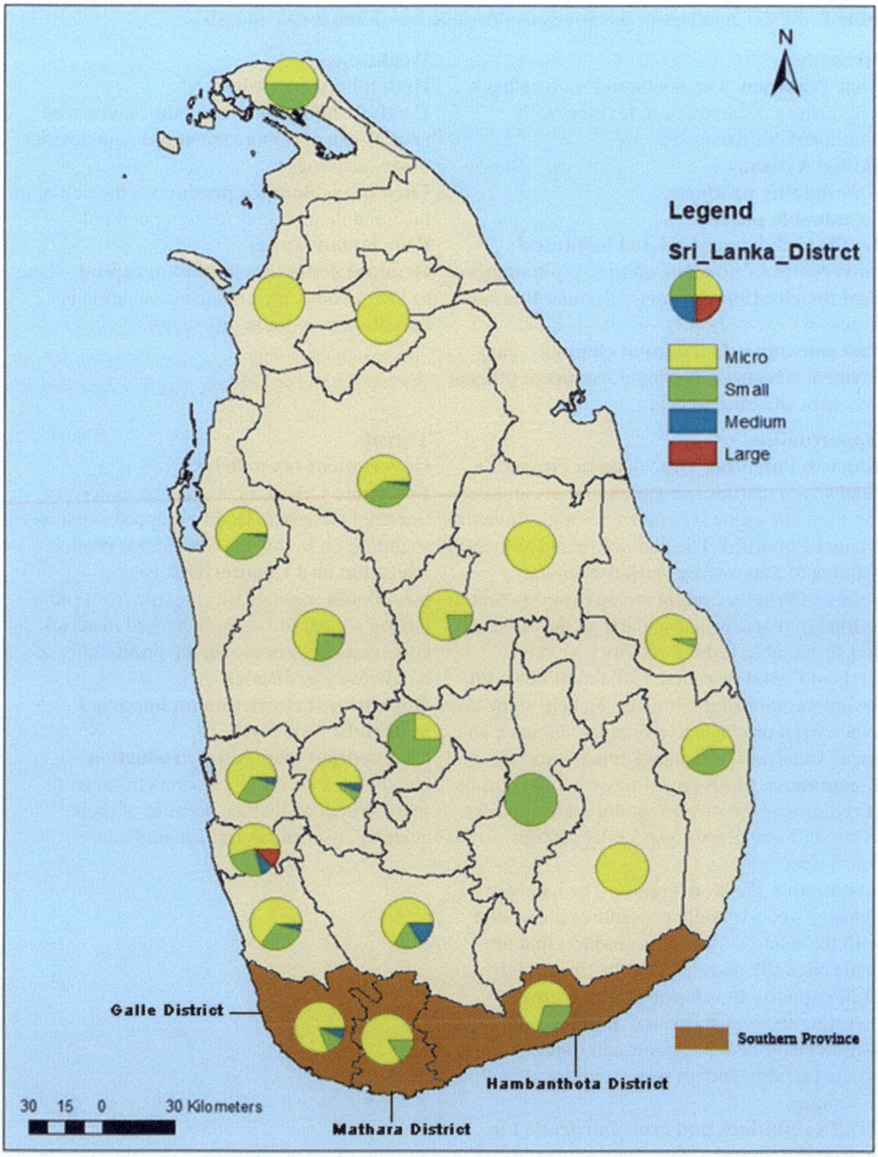

Fig. 1 Major Manufacturing Establishments of the Textile Sector by District in Sri Lanka. Aruna Banadara [6]. chrome-extension://efaidnbmnnnibpcajpcglclefindmkaj/https://www.industry.gov.lk/web/wp-content/uploads/2023/05/data-book-2022-Copy.pdf

Table 1 SWOT analysis of the Southern Province handloom textile industry

Strengths **Rich Tradition**: The Southern Province has a long history of handwoven textiles, with traditional craftsmanship **Skilled Artisans** **High-quality products** **Sustainable practice** **Textile and Apparel related institutes/ universities** Continuous training programmers **Vast distribution channel**—Ruhunu Ransalu stores **Vast consumer distribution channel**—easy payment schemes targeting government officers (teachers and other officers)	**Weaknesses** **High labour intensive** **Costly Production**: Producing handwoven textiles can cost more than producing textiles using machines **Lack of fashionable products**—the deficit of fashionable garments /design potential **Poor labour wages** **Gradual deduction in human capital**—Due to low income, the younger generation is unwilling to work in this sector
Opportunities **Tourism Potential**: The Southern Province's handwoven textiles can attract tourists looking for authentic cultural experiences and souvenirs **Export Potential**: The distinctive and authentic features of handwoven textiles from the Southern Province might create export potential to foreign markets, particularly in the artisanal and sustainable fashion sectors **Artisan Collaborations**: Collaborations with designers and fashion brands can help showcase handwoven products to a broader audience and create trendy, high-demand collections **E-commerce platforms**: Increasing the number of online sales platforms would make it easier to attract more clients, especially foreign consumers **Sustainable Fashion Trends**: The handwoven industry's eco-friendly procedures are in line with the rising demand for products that are made ethically and sustainably. (Natural dyes) **Skill/capacity development Programs**: Investing in local craftsmen' training and skill-development programs can improve product quality and preserve traditional techniques **Craft exhibitions and craft fair/trade fair** **Seasonal demand**—New year season (April month)	**Threats** **Government tax policies** **Economic Crises**: Economic recessions or financial crises can lead to reduced consumer spending on luxury and handmade products **Imitation and Counterfeits**: Export powerloom (poor quality) textile from India **Rising Costs**: Inflation, increased material costs can put pressure on the profitability of handwoven businesses **Government restriction on imported materials**: yarns/dyes **Competition from Mass Production**: Handwoven sector faces competition from mass-produced textiles because of their cheaper costs and quicker manufacture

Source By Authors

13 Data Collection Method

13.1 Primary Data Gathering: Participant Observation and Semi-Structured Interviews

The study used qualitative research techniques. Social scientists typically use qualitative research to create a detailed picture of how people behave and hold views in various contexts. Their desire to fully comprehend human behavior motivates them to engage in this discipline [93]. Along with linking people's thoughts and actions, this line of research also tries to investigate and characterize the "quality" and "nature" of how people act, experience, and comprehend through the use of non-numerical data [19].

Primary data collection includes participant observation and semi-structured interviews. The researchers conducted direct observation of departmental documents to gather more information. In order to learn more about the experiences, viewpoints, and expertise of handloom weavers, specialists, community members, and other stakeholders are interviewed. Multiple data sources enhance credibility by interviewing handloom weavers, specialists, and stakeholders and collecting secondary data from scholarly sources [82, 131]. Arvey and Campion [5] deem structured interviews more reliable than unstructured ones. Critics argue that the way questions are framed can influence the responses given.

Additionally, the interviewer's disposition may affect how the interviewee's comments are interpreted [5]. Interviews are the most common way of gathering information in qualitative data [110]. According to Rubin and Rubin [94] and Baxter [8], qualitative interviewing is a crucial method for data collection because it allows researchers to obtain rich and complex information from participants through meaningful dialogues and survey data for a comprehensive understanding of a phenomenon, according to studies [48, 60, 84, 93]. The semi-structured interviewing technique successfully promotes reciprocity between the interviewer and participant and enables the interviewer to improvise follow-up questions in response to participant comments. The well-liked semi-structured interviewing technique balances the interests of the researcher and the subject, allowing for candid responses and the emergence of themes [37, 99].

Interviews deal with thinking and talk later transformed into texts [92]. According to Crabtree [29], cited in [38], interviewing involves shared cultural knowledge. Seidman [97] and Denzin [36] presented differing perspectives on conducting interviews. They asserted that many favor interviews as a primary form of communication because of their intrinsic characteristics. As Seidman [97] puts it, "I interview because I am captivated by others' narratives." Through recounting stories, individuals convey fragments of their journeys.

Nevertheless, Denzin [36] contends that interviews go beyond mere information collection,they serve as mediums to craft performance texts and generate ethnographies that reflect individual and societal dynamics. Researchers gathered secondary data for this study from scholarly books and journal articles. Leveraging multiple data sources enhances the credibility of the data.

13.2 Expert Interviews

Expert interviews played a pivotal role in gathering data. Expert interviews comprise fourteen participants the Provincial Director, Weaving Instructors, Weavers, and District Officials from the Ministry of Industry of Southern Province. Human social existence is fundamentally based on verbal communication and face-to-face discussion to share our experiences and viewpoints of the world around us [99]. Interviewing, which has a solid foundation in social conventions of conversation, provides an accessible and efficient method for accessing the socially situated life worlds of individuals and groups and capturing a wealth of information on our informants' unique viewpoints and perspectives [14]. Conversely, most individuals agree that interviews take time since the researcher must go through a protracted process that begins with obtaining access to participants, continues contacting them, and ends with the interview [38, 91, 97]. At the beginning of the interviews, participants were informed about the research and received their consent about the research purpose. After the researcher interviews, analysis takes place. The analysis stage is challenging because of the enormous amount of data that was gathered during focus groups [26]. Transcribing spoken data to written data is the first step in the process. They made the questions simple, straightforward, and orally conducted. By the time of asking the questions from the experts, the method of breaking the interview into the five phases suggested by Robson [91].

1. Introduction: The interviewer gives a brief biography and discusses the purpose of the conversation.
2. Warm-up: Start with simple questions to make the setting more comfortable.
3. Main body: The interviewer concentrates on the primary subject of his or her research.
4. Cool-off: One more round of straightforward questions to wrap up the interview.
5. Concluding remarks: The analysis follows the researcher's interviews.

According to Robson [91], interviewers should refrain from asking detailed questions because doing so can force the interviewee to comprehend the entire question. Additionally, he advises against employing multiple- or double-barreled questions throughout the interview. It will be difficult for the interviewee to provide accurate answers to queries of this sort. He claims that using jargon in interviews is also frowned upon. Robson also encourages avoiding asking questions that steer the interview in a particular direction. He suggests that interviewers should remain unbiased and avoid posing any biased or leading questions. Oakley [79] claims that the interviewee responds passively and that the interviewer fosters rapport. Interviews were captured on tape and written down. Bryman [21] states three benefits of transcription: it preserves the interviewee's remarks. Interviews are crucial for qualitative research as they provide deep insights [18]. Interview participants were chosen based on their position within the institution. Moreover, rigorous protocols were in place during the interview to guarantee that researchers remained unbiased and maintained consistency across all sessions. The interviews aimed to identify opportunities for skill development, institutional motivation, opportunities for participation in training sessions for developing skills and learning, and education to fulfill active participation in the living heritage and living skills to cater to customer demand for present-day needs and socio-context.

Aim and Objectives: To identify how handloom textile weaving heritage prevails as a living entity and living skills in the Southern Province of Sri Lanka toward advancing Sustainability Goals 01 and 04.

Objectives of the study

To identify the characteristics of the living entity and living skills of handloom weaving as a heritage.

To identify the measures that have been taken to advance Sustainability Goals 01 and 04 (Table 2).

Researchers chose fourteen participants from both genders for interviews. Examine a series of questions asked of those working in the industry. They presumed all participants know the institution's guidelines and opportunities for skill enhancement, capacity building, learning, and education. In February and March 2023, researchers conducted fourteen interviews with Weavers, the Provincial Director, two Assistant Directors, the Deputy Director, the Director of the Textile Department, the Ministry of Industry, the Coordinating Officers, and Weaving Instructors

Table 2 Research proposition

Aim of the study	Objectives of the study	Research questions	Sources of data
To identify how handloom textile weaving heritage prevails as a living entity and living skills toward advancing Sustainability Goals 01 and 04	To identify the characteristics of the living entity and living skills of handloom weaving as a heritage	How many workforces of weavers, designers, and other officials strengthen the production of handloom weaving in Southern Province?	Primary data-interviews Secondary data Government reports Research papers
	To identify the measures taken to advance Sustainability Goals 01 and 04	In what ways are the strength of weavers, production levels, and empowered handloom textile weaving as a living entity?	Primary data-interviews
		How did the Textile Department empower them, and what rewards and incentives did they receive?	Primary data-interviews

who are familiar with government strategies, circulars, and institutional administrative processes. The industry aims to enhance the skill development and knowledge of the respective people through attending workshops and training sessions. Table 3 illustrates the summary of the interviews. Textile weavers continue working by using traditional skills, and the contemporary market seeks a new design orientation geared toward acquiring new knowledge. In structured interviews, interviewers use pre-planned questions that are organized and based on specific themes. These questions are designed to encourage detailed answers, guiding the discussion toward areas of interest. The format allows interviewers to delve deep into key topics they intend to explore during the session.

14 Achieving Sustainable Goal 1: Develop People's Skills for a Better and Wealthier Life

14.1 Literature Review- Traditional Skills and Craftsmanship

The skills are fundamental to securing decent jobs, and they are insufficient. It is crucial to link skills with employment opportunities and ensure they align with decent work standards. Additionally, incorporating skills as an integral component of economic growth and employment strategies is essential [55]. A growth strategy based on better education and training standards and accessibility can lead to a "virtuous circle" in which the improvement of skills fosters the innovation, productivity growth, enterprise development, technological change, and competitiveness required to sustain and accelerate the creation of new, better jobs [56].

The skills of people have been discussed in theories of craftsmanship by sociologists. They demonstrated that the skills of craftsmanship defined by comprehensive understanding, unstructured learning, intrinsic motivation, implicit knowledge, and problem-finding and problem-solving, which are both elements of the same process [9, 10, 41, 49, 74] [1952]; [98, 122]. Skills development can refer to any education and training geared toward the job, including apprenticeships and other shorter training programs. It can also apply to occupational education and workplace-based training programs. Skills of organization of craftsmanship built on informal networks and power derived from knowledge and experience and nested in a group of people with a similar task-oriented concentration. "Craftsmen" are other people whose abilities provide support in some way. Their labor is called "craft" [10].

Artisans embrace a second aesthetic requirement in addition to functionality: virtuosic talent. One who has mastered the techniques—an expert—has excellent control over the tools of the trade, can create anything with them, works quickly and nimbly, and can easily accomplish tasks that ordinary, less skilled artisans would find challenging or impossible [10]. The ability to perform tasks better than most others and more frequently is a component of virtuosity in some cases. Virtuoso artisans enjoy and praise their abilities within their trade and occasionally by outsiders [10].

Table 3 Interview summary

Code	Interviewee details	Purpose of interview
Interviewee 1	Former Assistant Director, Mathara District	Strategic plan, awareness of SGD
Interviewee 2	Provincial Director	Annual Development Plan Capacity Development Training Program SGD Design experimentation Infrastructure development Partnerships to enhance skills Innovations Administrative regulation Daily payments for weavers
Interviewee 3	Coordinating Officer, Galle District	Training and workshops
Interviewee 4	Deputy Director, Mathara District	Absenteeism of the weavers
Interviewee 5	Award-winning Instructor + Weaver + Designer, Mathara District	Skills of colour matching, Skills of identifying modem customers, Receiving motivations, Workshop experiences
Interviewee 6	Award-winning Instructor + Weaver + Designer, Hambanthota, District	Experience of learning skills
Interviewee 7	Assistant Director, Hambanthota District	Feedback from the panel of judges of the National Textile Competition
Interviewee 8	Coordinating Officer, Mathara District	Motivate district winners
Interviewee 9	Award Winning Instructor + Weaver + Designer, Galle District	Feedback as motivation
Interviewee 10	Director, Ministry of Industries, Textile Department	Value of feedback
Interviewee 11	Assistant Director, Galle District	Skills development
Interviewee 12	Teacher of Textile School	Training in textile schools
Interviewee 13	Weaver, Galle District, 2023	Workshops for new knowledge
Interviewee 14	Weaving Instructor + Weaver, Mathara District	Workshops for new knowledge, University premises enhanced knowledge

Some craftsmen refer to themselves not simply as craftsmen but as artist-craftsmen (a modern list might include potters, weavers, glassblowers, and furniture makers, to just the most prominent cases). Handloom textiles are identified as craft-based textiles that are being done exclusively by hand. According to Bourne et al. [13], craftsmanship embodies "the human condition of being engaged" in the task.

The craftsmanship theory is also context-bound, assuming a specific workshop-like social situation with informal connections and authority based on knowledge and experience. Harper [49] uses "his many-sided knowledge" of materials and his comprehensive grasp of engines and other devices to fix them and look for innovative ways to address problems in the real world. He can fix mechanical items by redesigning them because of his comprehensive understanding of engines and mechanical devices. Designing, producing, enhancing, and repairing are all interrelated. The social structure and Culture of the medieval workshop provide the perfect model for the craftsmanship method. The circumstance is quite different today. The emphasis on instrumentality and results and modern standardizing techniques have mingled with some aspects of the handicraft tradition [112]. The highly specialized skills are acquired through months or years of careful, prolonged practice aided by social assistance [42]. Even when adult artisans relocate, these skills will retain the particularity of the social setting they taught [42]. Mechanical and informational connections between a person and their physical environment define specific skills as behaviors [125]. The learning phase, during which the person explores the given area to identify the sensorimotor information required to complete the task, is crucial to the versatility of motor abilities [58]. Importantly, developing a skill is not a haphazard process but takes place in a given socio-historical context [16, 17]. Social coaching aids motor learning by directly interacting with older experts in the activity or structuring the learner's practice. This approach directs the learner's attention to their exploratory activity within an optimal section of the task space. Social guidance speeds Learning and creates a foundation for acquiring culturally unique skills [15, 16].

Further, Ingold [57] states that skill transmission influences three constraints: cultural influence constraints, task space constraints, and individual constraints. Social guidance speeds Learning and creates a foundation for acquiring culturally unique skills [15, 87]. The neuromuscular redundancy of the human body makes up the functional sensorimotor domain, and then several equivalents can accomplish task movement skills. Each society's "style" of skill represents a subset inside the available movement space, and it is this latitude in the execution of movement abilities that leads to a patterning of cultural heterogeneity [11].

Knowledge is the source of facilitates skills in human beings. In all human cultures, disseminating and transferring knowledge across generations is crucial. How people learn, what they learn, and who they learn from all contribute to who they are. Local societies have preserved traditions over time, so cultural expressions of traditional types of irrational knowledge have evolved. Theoretical perspectives on Learning and historical records detailing Learning give context for historians interested in how crafts learning took place in the past. The traditional domestic handloom textile industry-related education is one of the main streams that drive Sri Lanka's textile education. Education broadens the nation's capabilities concerning the moment's demands while supplying it with knowledge, skills, and procedures [28]. The schooling taught numerous improvements in the two main streams of sustainable survival.

ILO [55] states the importance of enhancing systems for predicting (current and future) skill requirements and sharing data to guide policy decisions, and training program design is needed [55].

> *"We have (Southern Province) strategic plan for every five years. This year, we made a Strategic Plan aligned to achieve SDG. According to the plan, we are setting up workshops and seminars to develop the skills of weavers and coordinators."*
>
> *(Interviewee1, Former Assistant Director, Mathara District, 2023)*

Table 4 illustrates that the strategic plan offers a clear vision and direction, laying out the measures required to realize the SDGs and ensuring everyone works toward a common objective. Further, Strategic Planning enables effective resource optimization and allocation for prioritizing the areas that demand the most attention and resources, which helps to avoid wasting effort. ILO [56] identified six advantages of bringing up a Strategic Plan,

1. To bring coherence to the system.
2. To streamline coordinated efforts and initiate reforms.
3. To aid in the coherence and coordination of policies.
4. To make institutional arrangements more clear.
5. To solidify current best practices.
6. To express political intent and commitment on behalf of all parties.

Recommendation No. 195 of the [55] offers policy recommendations for lifelong Learning, education, and training. Recommendation No. 195 posits that the government should primarily oversee pre-employment training. In contrast, employers should manage ongoing training, with individual workers taking the initiative to capitalize on educational and training opportunities.

> *"We have an Annual Development Plan for conduct projects which the funds allocated by the Provincial Council, so we were able to conduct relevant Capacity Development Training Programs for skills development of weavers, officers, and coordinators; with the economic crisis this year is tough for all of us, but previous years we successfully conducted Capacity Development Programs. According to the plan in 2024, we plan to achieve SDGs 8,9 and 10. We successfully achieved SDG 1 and 4 during the last four years."*
>
> *(Interviewee 2, Provincial Director,2023)*

Regular training programs ensure weavers consistently hone and develop their abilities, enabling them to produce higher caliber work over time. There are five advantages, such as using new methods, ensuring quality, boosting self-assurance, promoting economic growth, and maintaining tradition, which are vital advantages achieved by weavers and holistically by the Textile Department.

> *"We developed a Strategic Plan to meet the SGD within 2020-2024, and now we are actively engaged in proceeding with it. We identified SGD 1 and SGD 4 as essential to meet SGD 8,9 and 10. I can see a correlation between SGD1,4,8,9, and the core point is skills development and Learning."*
>
> *(Interviewee 2, Provincial Director,2023)*

Table 4 Strategic Plan 2024

Southern Province Strategic Plan—2024					
Purpose	Aim	Sustainable development goals	Goals	Strategies	Performance indicator
1. Becoming the main provincial contributor to the national handloom production through the handloom industry	(i) Increas annual production	8, 9, 10	*50	Increase investment in machinery repair and effectively contribute to production	Number of machines added
		8, 9, 10	* 50		New Manufacturing company
		8, 9, 10	30	Adopting strategies to attract new workers to manufacturing operations	Number of workers engaged in production
		8, 9, 10	10%	Planning of production operations including market stratergies	Increased Industrial Wages (Rs.)
		8, 9, 10	5% Reduction	Designing the production process to attract workers	Percentage of work leavers
		8, 9, 10	Annualy 40,000 kg	Developing capacity	Amount (Kilograms) of the dyed yarn
	(ii) Improving the quality of products	8, 9, 10	Minimum hours 30	Development and updating of knowledge in field	Number of training hours
		8, 9, 10	50		Number of workers trained
		8, 9, 10	Reduction of annual production capacity by 50%	Develop quality control checking system	Amount of goods in failer condition
	(iii) Increasing annual income	8, 9, 10	22	Introducing marketing team	Number of Marketing Managers
		8, 9, 10	02		Number of new showrooms and outlets

(continued)

Table 4 (continued)

Southern Province Strategic Plan—2024					
Purpose	Aim	Sustainable development goals	Goals	Strategies	Performance indicator
		8, 9, 10	10%	Creating new marketing opportunities	Increased sales revenue
		8, 9, 10	5500	Promotion programs among customers	Increased customer base
		8, 9, 10	60%	Strengthening the debt recovery process	The amount charged per annum
		8, 9, 10	Reduction of transaction time to 05 min		Transaction time

"Even though Sri Lanka is facing economic difficulties, we, as the management, are readily available to conduct strategies aligned to SGD; we are much aware of these goals and responsible for achieving them."

(Interviewee 1, Former Assistant Director, Mathara District, 2023)

It is vital to enhancing mechanisms to forecast current and future skills needs and distributing this information is vital for guiding policy formulation and shaping training plans [55].

"We are in the latter part of the process of opening a Design Center in the Galle District; according to the plan, this is where any designer or weaver can do any design experimentation before beginning their design. Previously, we had no place to explore and weave ideas in samples. A long time back, the University lecturers taught us to experiment as much as possible. So, now, as a Province, we are much delighted to have a Center like this."

(Interviewee 2, Provincial Director, 2023)

The Textile Department of Southern Province realized the vacuum of textile design experimentation. To fill the gap, setting up a design training institution is an essential hub for promoting originality, innovation, and valuable skills. Their significance for new experimentation fosters five benefits: encouraging creativity, providing access to resources, fostering collaboration, improving skills, bridging theory and practice, and fostering industry links.

Gaining decent jobs requires skills that are essential but insufficient on their own. Relating skills to employment prospects and decent work is crucial, and employment and economic growth initiatives must consider skills [55].

'According to the Government Circular, each officer must undergo a Training Program annually. So, we are following this guideline. We normally set these programs with the state and private sector partners to enhance their expected skills demanded by their job roles.'

(Interviewee 2, Provincial Director, 2023)

'We currently have a partnership with an entrepreneur who develops yarns from Banana fiber; we successfully experimented with weaving using Banana fiber. We aim to use banana fiber for future designs, so we will be able to reduce the use of cotton fiber for weaving and will be able to save some amount of money. We think of utilizing the saved money to increase the number of training programs.

(Interviewee 2, Provincial Director, 2023)

'I received complete training to work in a dye house. I participated in many workshops. I learned new knowledge from those workshops, seminars, and training. The past two years have been critical for us; we hope in the next year, the situation will become stable.'

(Interviewee 3, Coordinating Officer, Galle District, 2023)

'We operated twenty-four weaving centers and are now operating only twenty-two. We had nearly one hundred twenty workforce weavers during the past six months. Attending to the work of weavers is irregular due to their private reasons. We have sumptuous looms in three districts. We paid weavers daily according to the pieces they completed. We pay different payments for each activity they engaged in for each task, for example, bobbin winding, warp preparation, warp beam preparation, Attaching warp yarns on the loom, weft yarn winding, and weaving'.Skillful weavers earn a lot.'

(Interviewee 1, Former Assistant Director, Mathara District, 2023)

Figure 2 illustrates the availability of handloom machinery in the Southern Province. Galle district occupies more looms than the other two districts.

Prolonged engagement with a particular task enlightens the person who carries work. Weavers can repeat procedures to continuous interaction, and repetition is the key to mastering any skill. They improve their skill at their craft through time. Sometimes, a longer commitment exposes weavers to more difficulties. By tackling these problems, they can improve their problem-solving abilities and promote innovation. Longer durations of weaving enable individuals to develop a deeper understanding of the components, processes, and subtleties of design, resulting in more intricate

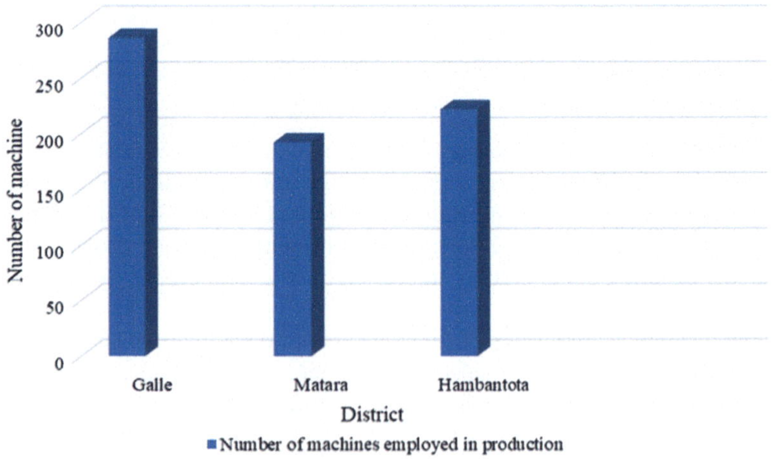

Fig. 2 Availability of handloom machines in southern province

patterns and textures. Weavers develop muscle memory through consistent practice. As hands get used to the motions, weaving becomes more intuitive and straightforward. Figures 3, 4, 5, 6, 7, 8 and 9 show the weaving process and illustrate the tasks of completing the weaving procedure, which requires skills. Spending more time weaving boosts confidence. Weavers gain increased confidence in their abilities with every successful project, prompting them to undertake more complex tasks. In essence, sustained weaving practice improves a craftsperson's technical abilities while strengthening their ties to the craft's tradition and market realities.

The process of weaving

> *"Impact of the current economic regression: nearly one hundred weavers left the job. Early in our workforce, there are more than six hundred in the Province. Retaining the weavers is critical at this moment. At present, we have 530 workforces. Our weavers are our strength. We hope to recruit new weavers from new generations by adopting new strategies. We have to wait patiently".*
>
> *Interviewee 4, Deputy Director, Mathara District,2023)*

The Deputy Director highlighted the importance of a thoughtful mechanism for maintaining the present workforce, as shown in Table 5. They forward some initiatives for approval from the Ministry. Figure 10 portrays the annual budget allocations for skill development and capacity development of weavers, designers, and coordinating officers of the Southern Province.

> *'Normally, we allocate a budget for workshops and skills and capacity development of weavers, instructors, designers, and coordinating officers.' We were able to dispose of the allocated budget meaningfully every year.'*

Fig. 3 Bobbin Winding

Fig. 4 Warp preparation

Fig. 5 Warper

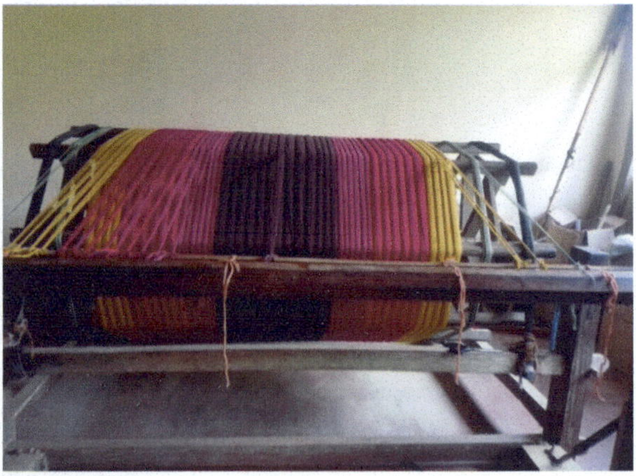

Fig. 6 Warp beam preparation

Fig. 7 Picking

Fig. 8 Shuttle is kept on the fell of the cloth

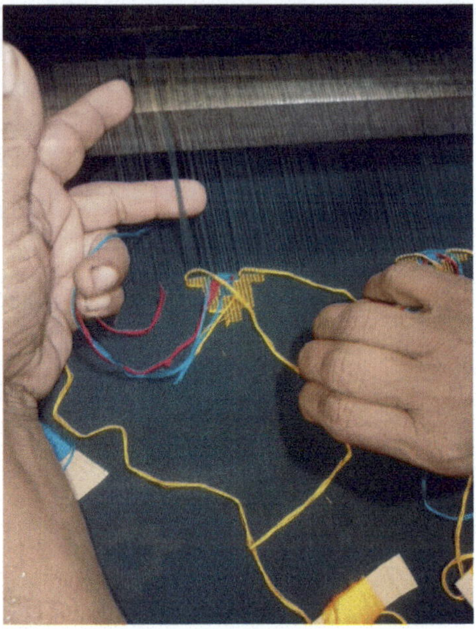

Fig. 9 Weft insertion for a design ("*mal bama*" technique)

Table 5 Present workforce at the textiles department in the southern province

District	Weaving centers	Dye centers	Assistant directors	Weaving instructors	Weavers
Galle	32	1	1	32	210
Mathara	21	1	1	32	120
Hambanthota	25	1	1	32	200

(Interviewee 2, Provincial Director, 2023)

Groth [45] and Huotilainen et al. [54] highlighted the significance of crafts in education due to their influence on our embodied cognition and their provision of a hands-on learning environment. Such an environment helps integrate abstract learning concepts and skills.

> *'We learned skills of color integration in the series of training programs with modern design concepts that demand current handwoven clothing preferred customers. I think the traditional skills of color matching can be mixed match and forward to modern consumers who prefer both traditional and modern.'*
>
> *(Interviewee 5, Award-Winning Instructor + Weaver + Designer, Mathara District,2023)*
>
> *'After learning new skills in design development, I integrated that knowledge to produce sarees. Our Coordinator appreciated those designs. I received an award for my creations, so I'm happy. I'm ready to learn new skills for design'.*
>
> *(Interviewee 6, Award-Winning Instructor + Weaver + Designer, Hambanthota District, 2023)*

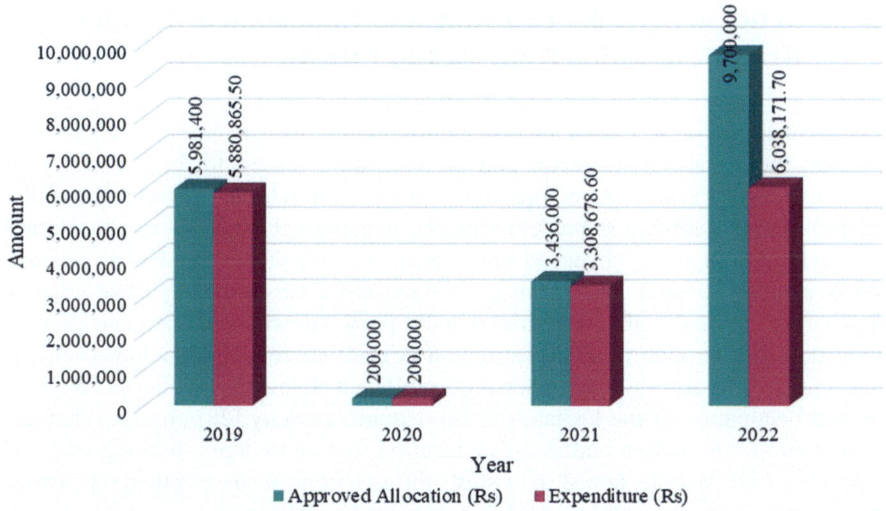

Fig. 10 Annual budget allocations and expenditure for skill development

Weavers can routinely practice and refine their craft on weaving looms, which eventually leads to proficiency. They augmentation of skills as they use the loom more frequently.

"The government provides us with a handloom machine and thread to work from home after receiving instruction in the training facility."

(Interviewee 6, Award-Winning Instructor + weaver + Designer, Hambanthota District,2023)

Implementing a systematic Strategic Plan for the textile industry that aligns with the national SDG policy is of timely importance. The current Strategic Plan of the Southern Province aims to coordinate and facilitate transformative measures to strengthen the Southern Province (Table 4). This plan outlines essential performance indicators to realize the vision of the skills system cohesively. This strategy fosters a more holistic approach to human resource planning, bringing together various government agencies and education and training providers to onboard the next generation, propelling the industry's future.

A revitalized textile sector requires diverse skills and capabilities from individuals across various socioeconomic and demographic backgrounds. Short-term courses, adaptive learning, and an amalgamation of creative and entrepreneurial acumen highlight the growing importance of lifelong Learning as a critical career driver. Incorporating sustainability-centered creativity, business, and entrepreneurship principles into curricula across multiple academic disciplines is vital. Educational frameworks should integrate the expertise and insights of Sri Lankan designers and business owners actively involved in the sustainable textile industry.

15 Practice of SGD 04: Learning and Education of Textiles: A Brief Description from Past to Present

Textile education in Sri Lanka navigates through two main avenues: the traditional domestic handloom textile sector and the education related to the export-driven apparel industry. Education equips the country with relevant knowledge, skills, and techniques, evolving in tandem with the demands of time. These educational pursuits have brought significant advancements, ensuring the sustainable progression of both primary streams. In Sri Lanka, technical skills, encompassing knowledge of engineering, art, and crafts, were traditionally passed down from one generation to the next, either through familial instruction or craft apprenticeships. However, the conventional education system underwent significant changes following the Western colonial occupation of the lowlands [128]. Coomaraswamy [28] observes that the foundation of a Sinhalese craftsman's education was an in-depth, lifelong study of a specific craft. Artisans honed their skills through continuous practice, repeatedly copying and mastering patterns until they achieved excellence.

Further, Coomaraswamy [27] notes that operating looms in Sri Lanka closely mirrored techniques prevalent across Asia. Typically, the loom was situated in an open shed within the weaver's residence. Consequently, the weaving tradition in ancient Sri Lanka is perpetuated across generations with significant advancements.

Under British rule, Sri Lanka experienced a notable quantitative growth in education. Vernacular schools established in villages emphasized skills-based vocational training. The initial phase of British governance in Sri Lanka, from 1796 to 1948, brought about significant transformations. In 1859 the beginning, missionaries established over 100 industrial schools across the island. These institutions initially provided underprivileged youth with training in agriculture and crafts. Once these schools came under state ownership, their focus gradually shifted. They began training students under the Department of Education and, subsequently, the Department of Industries, preparing them for roles in the burgeoning public industries [111].

In 1944, Education Minister CWW Kannangara pioneered Sri Lanka's free education system, shifting the educational landscape from a privileged English-educated elite to broader accessibility, emphasizing national languages and promoting vocational training over industrial schools, with weaving introduced for the underprivileged, all underpinned by his belief in experiential Learning and the holistic development of skills, emotions, and intellect. During Kannangara's era, weaving schools, including prominent ones at Horethuduwa and Hiramadagama, were established. They founded the Central Weaving Institute in 1936. Recognizing textiles as a pivotal domestic industry, the Council of Industry Commission endorsed yarn production, textile weaving, and T-shirt crafting. 1944, 125 textile training centers were established under the Ministry of Industries and Commerce. However, they identified caste-based traditions in art and craft as barriers to growth. In response to these barriers, they designed 6-month technical courses for 16–18-year-olds and recommended establishing training centers in villages centered around crafts. In 1947, weaving was incorporated into the Technical College curriculum [128]. Established in 1985, the Department of Textile and Apparel Engineering at the University of Moratuwa initiated the BSc Honours Degree Program in Textile Engineering, which benefitted from collaboration with the Textile Department of the University of Leeds and Manchester Metropolitan University. By 2001, in partnership with the London College of Fashion from the University of Arts London, an expert in fashion design education, the Department introduced the Bachelor of Fashion Design and Product Development Honours Degree course with two pathways: fashion design and textile design. University of Moratuwa, Sri Lanka, is the country's pioneering education provider of Textile Education in textiles' design and engineering aspects. Chart 1 shows the evolution of the Textile Education System in Sri Lanka.

16 Informal Knowledge Sharing: After the Provincial Textile Competition

The handloom textile industry in Sri Lanka also relies on vital informal knowledge sharing, predominantly seen at the Provincial Textile Competitions organized by the Sri Lankan government. The government regularly conducts these national contests to promote and support handloom textile weavers. Sri Lanka boasts a rich history of handloom textile production. To encourage innovative designs and high-quality production, the government hosts National competitions and arranges Provincial textile exhibitions nationwide. The pinnacle of these events is the Annual National Exhibition, designed to capture the interest of foreign consumers and buyers. This exhibition is a testament to local weavers' genuine creativity and expertise.

'During the last few years, the Galle district received the golden award, and we got the most awards in the Southern Province.'

(Interviewee 2, Provincial Director, 2023)

In the Provincial competition, the weavers, designers, and officials receive constructive feedback from the panel of judges regarding design orientation, color combination, the technicality of the product, and market viability to eradicate weak points of their product.

'Most of the weavers participate in the annual competition. They are eagerly waiting to submit their new creation annually for the competition. The most valuable point is that competitors received comments to develop their products for the following year. They also receive the reasons for rejecting their product. So, they did not have any doubts about their product selection criteria.

(Interviewee 7, Assistant Director, Hambanthota District, 2023)

Offering monetary rewards as a form of motivation taps into a fundamental aspect of human behavior: the desire for tangible benefits. For many years, monetary incentives have been researched and discussed in psychology, business, and economics. Offering monetary rewards can serve as a compelling motivation tool in various

Chart 1 Evolution of the textile education system in Sri Lanka

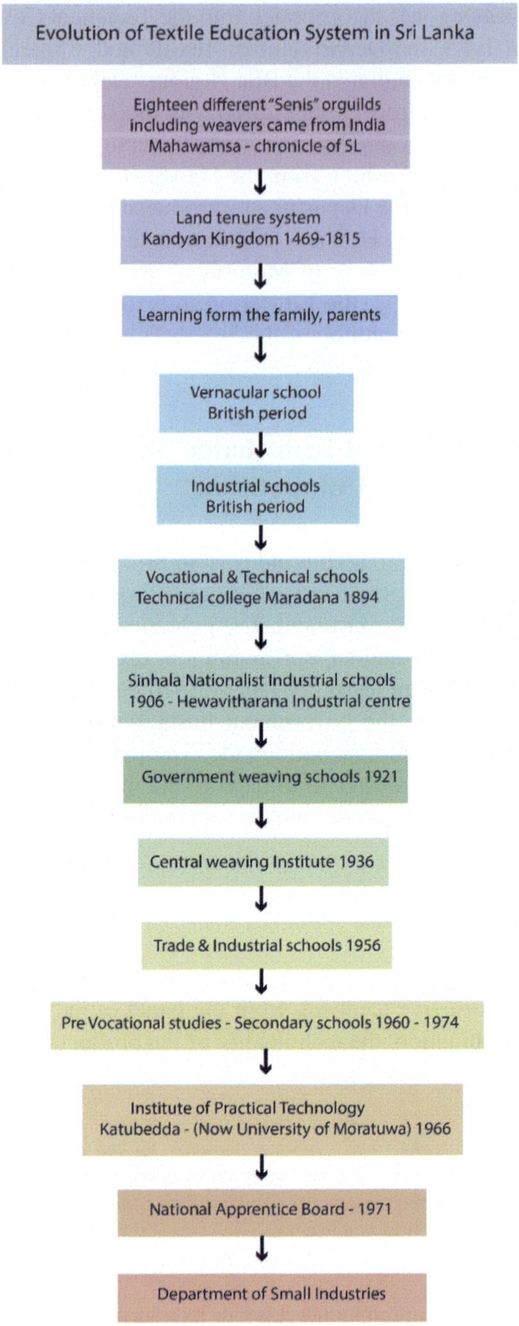

contexts. Scholars have identified seven factors that impact the monetary prize to the receiver.

1. Immediate Gratification: Financial incentives provide an instant sense of reward and satisfaction. This immediate feedback mechanism can strongly reinforce behavior or action [67].
2. Flexibility of Use: Monetary rewards allow recipients to choose how they use their prize, whether for personal desires, essential needs, or savings [24].
3. Quantifiable Achievement: Financial incentives can directly measure success, providing a clear and tangible gauge of performance [66].
4. Competitive Drive: Monetary rewards, especially in contests or competitions, can fuel an individual's competitive nature, leading them to outperform others for the prize [22].
5. Tangible Recognition: Beyond verbal appreciation, monetary rewards offer a concrete recognition of one's effort, making individuals feel valued [43].
6. Goal-Oriented Behavior: Clear financial rewards can set precise targets for individuals, pushing them to achieve or surpass specific benchmarks [95].
7. Improved Value Perception: Financial rewards can elevate the apparent worth of tasks that may appear ordinary, spurring significant effort or engagement [4].

However, researchers also emphasize the complexity of human motivation and the potential downsides of over-relying on monetary incentives. For instance, [34] found that extrinsic rewards, like money, can sometimes reduce intrinsic motivation. It is essential to understand when and how to use monetary incentives effectively, considering their potential impact on other motivational factors.

'I won the Golden Award for best sari for five consecutive years. I received plaques and a cash prize, too. I think I anticipate both types of valuables. Cash prizes naturally motivate humans, I guess.'

(Interviewee 5, Award-Winning Instructor + weaver + Designer, Mathara District, 2023)

Economist Mariana [73] asserts that in a society where "value equates to price," any activity with a price tag is perceived as value-creating.

'Usually, as a Province, we offer cash prizes for national-level textile and Provincial winners. Recently we could not offer such a prize due to an economic crisis.

(Interviewee 8, Coordinating Officer, Mathara District, 2023)

On-the-spot feedback, when delivered effectively, can significantly motivate individuals. Immediate feedback allows individuals to promptly recognize what they did correctly or where they might need to adjust. The immediacy of the feedback enhances its impact.

1. Instant Correction and Reinforcement: Immediate feedback allows instant correction, leading to quicker behavioral adjustments and faster Learning [51]
2. Increased Engagement: On-the-spot feedback can keep individuals engaged and prevent them from continuing down an ineffective or incorrect path [103].
3. Validation and Recognition: Prompt feedback makes individuals feel seen and valued, enhancing intrinsic motivation [108].

4. Enhanced Learning: Real-time feedback aids in the learning process by helping individuals immediately understand and internalize the right actions or methods [129].
5. Building Trust: Immediate and constructive feedback can foster trust, as it indicates that supervisors or colleagues are paying attention and are invested in the individual's growth [39].
6. Encouraging Continuous Improvement: Continuous, on-the-spot feedback encourages a culture of ongoing improvement and adaptation rather than waiting for periodic reviews [33].

'I did not win any National level awards for my sari designs. I received Provincial awards for last year; I firmly believed that, based on the feedback received in the previous year, I would be able to produce a competitive product this year.'

(Interviewee 8, Coordinating Officer, Mathara, 2023)

'I have participated in this competition for more than ten years now. I won awards several times. I developed designs and considered the technical aspects of products based on the comments from the judges. On-the-spot feedback is useful, I think.'

(Interviewee 9, Award-Winning Instructor + Weaver + Designer, Galle District, 2023)

'In the National Judging, as the Director, I formally requested the National Panel of judges to give constructive feedback to the Provincial Directors and relevant authorities who represented their respective Provinces. So, all the time, judges give valuable comments to them. I aim to communicate the feedback to the weavers and designers who participated in the final round, who are not permitted to participate as observers at the final judging.'

(Interviewee 10, Director, Ministry of Industries, Textile Department, 2023)

'In the National Judging, as the Director of Galle District, I usually recorded the feedback, which is convenient for me when passing the comments to weavers and designers in my District. At that point, I tried to get clarification from the judges. They give beneficial ideas that weavers can adopt for the following year's competition.

(Interviewee 11, Assistant Director, Galle District, 2023)

The successful outcome of developing the skills of the weavers and the dedication toward producing creative and innovative handloom products for the Annual Textile Competitions is evident in Table 6. Southern Province won the championship for seven consecutive years and received a Gold Award for five years.

'We take weavers, designers, and coordinating officers for shop visits in Colombo to develop their skills in identifying modern-day consumer needs. We took them to Barefoot and Seylyn, handloom textile boutiques, several times, which produce traditional and modern design concepts to meet contemporary customer needs all the time.

(Interviewee 2, Provincial Director, 2023)

Few such handloom products are available in the current mass market due to the absence of weaving in such delicate and sophisticated woven structural patterns. The intricate weaving design is, therefore, reserved exclusively for competitions. Thus, there was a good market for these complex patterns. The Provincial Council hosts an exhibition after the Provincial competition concludes, encouraging other business partners to purchase their current items.

Table 6 Achievements of southern province in the national handloom competition in seven years

Year	2016	2017	2018	2019	2020	2021	2022
Overall championship	1	1	1	1	1	1	1
Golden award	1	1	1	1	1	–	–
Silver award	1	–	–	–	1	1	–
Bronze award	1	1	1	1	1	1	–
First place	8	8	10	12	7	5	11
Second places	8	3	4	4	9	12	10
Third place	3	5	7	5	3	6	7
Merit 01	7	6	5	5	5	7	8
Merit 02	3	7	7	3	3	5	7
Total marks	102	81	98	107	108	102	114

17 Pursuit of Academic Partnerships and Collaborations: Workshops

Since the 1990s, there has been an increasing understanding that higher education institutions (HEIs) must serve and be relevant to the communities and regions located, especially when taxpayers question the purpose of these government institutions [46]. Department of Textile and Apparel Engineering collaborated with the Textile Department of Southern Province and conducted a series of workshops for capacity building of weavers, designers, and officers of their districts, consecutively targeting the Annual Textile Exhibition, which is patronage by the government of Sri Lanka. These workshops set a social innovation-oriented suite of experiential learning tasks that enhanced creativity and identified the design orientation of contemporary needs of handloom textiles. Workshops highlighted the articulated interests and needs discovered within the weavers and designers of the Southern Province handloom textile groups guided by the concepts of action learning. Holland [52] states that higher education institutions must participate in a highly complex learning society where engagement, discovery, and learning are intertwined activities that draw on various sources of knowledge produced in various contexts by various individuals. Learning that fosters collaboration among numerous players and obfuscates the distinction between informal and formal Learning is especially needed [85, 124]. Universities should actively cultivate connections with the community by co-producing place-based knowledge and activities that support community sustainability and democracy. People should base these activities on the communities' needs they regularly interact with [68, 85].

Mtawa and Wangenge-Ouma [77] identified that instead of just imparting knowledge and services to the masses, the current focus is on fostering collaboration, mutual respect, and shared Learning between establishments, such as Higher Education Institutions (HEIs) and their respective communities.

Figures 11, 12, 13 and 14 showcase a sequence of design skill development programs conducted by the academics of Fashion Design and Product Development Degree courses in the Textile Department of the University of Moratuwa on invitation by the Textile Department of Southern Province in 2019. The primary aim of these in-house training sessions was to invigorate and inspire the weavers, designers, and coordinating officers to hone their design skills in alignment with market demands. The faculty discerned that the market favors traditional motifs and a fresh approach to design placements.

> *'I have been weaving for over fifteen years, but the design techniques I learned in the training were entirely new. I never realized my ability to sketch and create motifs. The training showed me a unique approach to design I had not encountered before. After attending those sessions, I discovered my innate design talent. Subsequently, I even won an award in the Annual Competition.'*
>
> *(Interviewee 9, Award-Winning Instructor + Weaver + Designer, Galle District, 2023)*

Ahern [3] highlights that by incorporating transdisciplinary knowledge into local sustainability plans, communities evolve into "test grounds" for urban sustainability, primed to implement effective, ecosystem-centric solutions. In socio-technical studies, they designate the term "niche" to crucial areas where they safeguard knowledge transfer and explore innovative ideas against market pressures. University-community partnerships can create these niches, breeding grounds for pioneering sustainability strategies. These spaces are essential for nurturing fresh concepts since broader society, with its set cultural norms and values, often resists change, potentially cementing unsustainable paths [23].

Weavers and designers of the Southern Province experienced the creativity and resourcefulness that offer a regenerative approach, enhancing the design capability at the workshop conducted by the University of Moratuwa a couple of years ago.

Fig. 11 Evaluating design work in the workshop held in 2019

Fig. 12 Development of inspirational skills; copying original designs

Fig. 13 Skills of scaling out motifs

Fig. 14 Teaching skills of drawing design mofifs

A series of workshops conducted in 2016–2020 enhanced the capacity of designing in terms of selecting yarns, colors, techniques, textures, and methods of handloom textile weaving in Southern Province.

The following domains of the workshop programs conducted,

(a) Competency-Based Design Training and Design Innovations
(b) Promote traditional design motifs and design composition to an eclectic mixture
(c) Develop color selection competency and color usage

The Provincial Director and Assistant Directors of three Districts with relevant officers involved participate in the workshops with weavers and designers. Abreu [1] valued collective action, which can involve economic and non-economic linkages, depending on the capacity to mobilize convergent interests through any agreement. Textile Department academics of the University of Moratuwa acted as solution providers for the issues about the design capabilities of the weavers and the designers. As a result, the handloom textile sector must adapt its business models for sustainability to produce benefits for the economy, society, and the environment and establish credibility for its objectives. Chart 2 provides the way forward road map to archive the Annual Competition.

The workshop's learning component is a way to boost confidence by relating the activities of organizations to the four pillars of prosperity to increase knowledge and comprehension. It is also an opportunity to confirm the accuracy of the implicit work by bringing fresh suggestions for what else can do, how achievements can be built upon, and gaps and possibilities found to advance the handwoven textile industry.

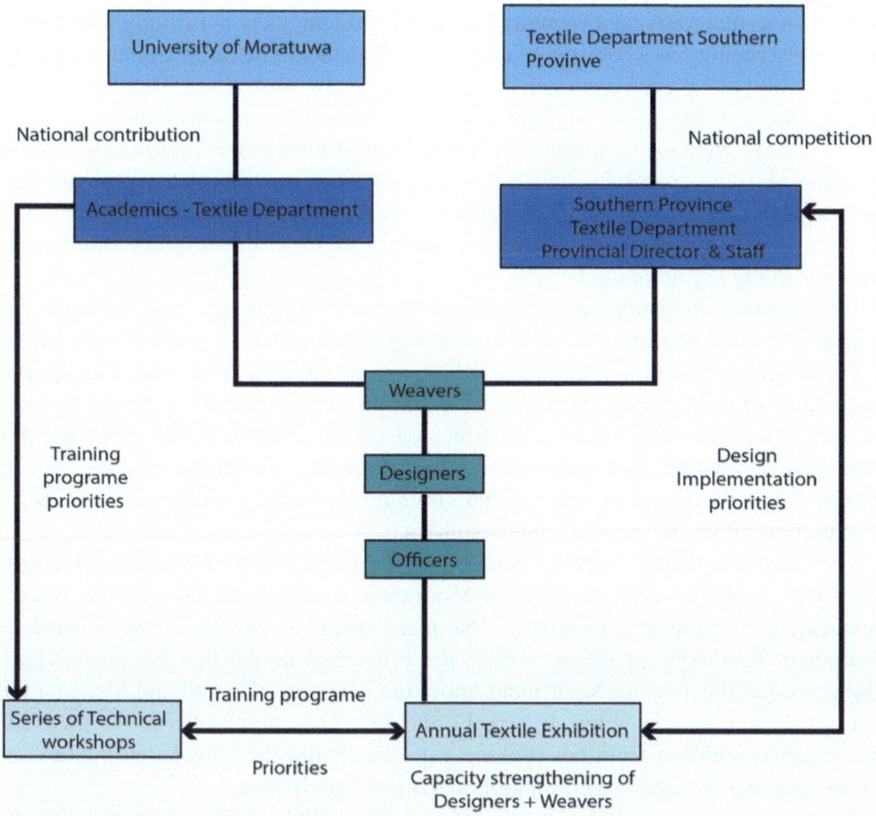

Chart 2 Pursuit of academic partnership and collaboration with southern province textile department

18 Conducting Practice-Led Learning Opportunities: Practice-Led Learning

The modification of practice-led research drives the concept of practice-led Learning. Craft education values practical training over establishing a theoretical and conceptual foundation. Textile education at the University of Sri Lanka often organized workshops for designers, technicians, and industry stakeholders to acquire technical skills. Physical characteristics, such as tensile strength and elasticity, are used to teach about the medium's materiality. Students can experiment with conceptual or expressive aspects in their studio work after thoroughly discussing the properties, such as feeling and impression. The creative process is sparked by skills, followed by selecting a material that complements the technique, or vice versa. An excessive focus on textile techniques can limit an artist's thinking and impact the choice of materials for each method. For instance, a student majoring in weaving might only consider using yarn as materials on a weaving loom, or a student adept at printing might only view silk-screening as a pattern on a fabric's flat surface. The range of materials that can be employed for each approach is thus significantly constrained. Therefore, when the weaving projects are given to the undergraduates, they can select any material in the section on weaving techniques [31, 30]. Southern Province Textile Department identified design capabilities as a deficit of handloom weavers. They realized that the design education and technical training sector is strategically vital in enhancing production growth because of the contemporary demand for designer handwoven products in the textile market.

Furthermore, they realized that design training bridges the gap between the general technical education offered in weaving schools and the specialized courses of Technical and Vocational Institutes, which are not part of the standard educational curriculum in the country. Therefore, the design of the education system is vital for Sri Lanka, given the high rate of educated youth in the rural areas who are not employed. Workshops and sessional training offered by Universities having Textile Design Degree Courses is an essential strategy for linking training and capacity development of textile weavers and designers.

The textile workshops were conducted by the Department of Integrated Design, Faculty of Architecture, University of Moratuwa, in collaboration with the Textile Department of Southern Province. The main resource person of the workshop conducted the theory of colors, which she pioneered as the first textile designer who served in the Textile Department under the Ministry of Small and Medium for more than twenty years (Figs. 15, 16, 17 and 18). The overall idea of the program was to get inspiration from her lifelong experience and the knowledge pursued in textile weaving at Camberwell School of Art in Great Britain.

Figure 17 shows that when presented with immediate feedback, many weavers and designers discover their untapped potential, seeing it as an opportunity to pause and reflect. This method highlights their progress and boosts their confidence, encouraging them to embrace the four pillars of Culture, Society, Environment, and Economics for a comprehensive approach to sustainability.

Fig. 15 Learning by doing, colour matching. Photo Credit for Department of Integrated Design, University of Moratuwa

Fig. 16 Learning by doing, colour theory. Photo Credit for Department of Integrated Design, University of Moratuwa

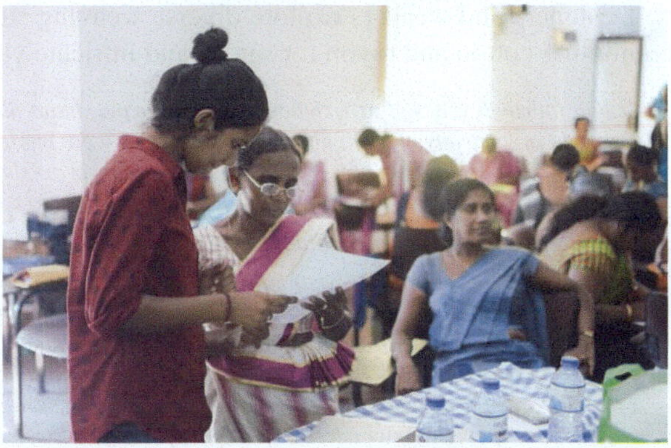

Fig. 17 Verifying tasks. Photo Credit for Department of Integrated Design, University of Moratuwa

Fig. 18 Designs for contemporary market. Photo Credit for Department of Integrated Design, University of Moratuwa

The resource persons of the workshop showcased the possibility of using hand-woven textiles in contemporary fashion (Fig. 18). Handloom textile design leadership, when combined with effective idea execution, is paramount. Recognizing the pivotal role of sustainable textile Micro and Small Enterprises in Sri Lanka is crucial. By tapping into the immense potential of these innovative entrepreneurs, they pave the way by establishing new market niches and exemplifying best practices. Such efforts enhance the textile industry in Sri Lanka and motivate peers to emulate these successful strategies.

'Every year, we host several seminars for students using the textile sector as an example. Additionally, those students visit our training facility where they gain extensive machine operation training.'

(Interviewee 12, Teacher of Textile School, 2023)

Designers and weavers explore diverse weaving structures, colors, and patterns, combining cotton and rayon for sarees and intricate yarn pairings.

'I had fourteen years of experience. For this carpet, hana fibres were used. I made the weft yarn out of a variety of hana fibres. These carpets were made to be strong, long-lasting, and beautiful. The know-how I received from the previous workshops conducted by the University lecturers was also precious. There I learned about how to use natural dyes to hana fibres."

(Interviewee 13, Weaver, Galle District, 2023)

"I was experienced—18 years. I studied textile weaving in a Textile School a long time ago. This award-winning tablecloth was made from a single skein of yarn I got this new knowledge from the workshop conducted by the University. Weft yarns of various colours were used to prepare this. I explored many other techniques as well. Internal Workshops conducted at the University are really valuable because we experience new knowledge in the real design setting, which is a golden opportunity for us.'

(Interviewee 13, Weaver, Galle District, 2023)

"We used beautiful colour yarn and rayon that the government had introduced for sarees.'

(Interviewee 13, Weaver, Galle District, 2023)

'We identify customer needs by participating in training programs often. We learned modern design concepts and now know how to combine traditional and modern design ideas. It is incredible!'

(Interviewee 13, Weaver, Galle District, 2023)

Using various dyeing techniques is also a draw for this sector. Weavers employed chemical dye to color the strands. It is frequently utilized right now as well. However, someone very creatively exploited it. They colored some portions with one hue and others with anthracite using rubber bands. The weft yarn is quite accustomed to it.

'In the last five years, we have increased the revenue; I think we managed the consistency of the workshops and training for weavers and coordinators. The impact of the workshops on the production is evident."

(Interviewee 2, Provincial Director, 2023)

Figures 19, 20 and 21, illustrate the growth in yearly earnings in the Southern Province. The net profit metrics reveal the Net Income, representing the industry's earnings. This metric helps the government to determine the industry's ability to yield returns on its investments and enhance its overall worth. Conversely, the comprehensive income statement displays variations in net equity.

> *"We used beautiful mixed different yarns that the workshops had introduced for sarees. We used fabric strips as an alternative for yarns. I think University workshops broaden our knowledge, and we lose our fear of using new types of yarn; there is a great demand for these new designs today.'*
>
> *(Interviewer 14, Weaving Instructor + Weaver, Mathara District, 2023)*

Figure 22 portrays the online sales revenue of the Southern Province handloom website. It is visible that there is a gradual decline in online purchases of products during the COVID-19 pandemic.

It is realized that workshops conducted by the University were successful based on the knowledge generated by the resource persons. The weavers, coordinators, and instructors were delighted about the new knowledge they received. The participants in the workshops valued continuous training and updating knowledge. Products should evolve into functional crafts, enhancing and magnifying users' skills, mindset, and capabilities to foster greater creativity and durability [130]. A robust market necessitates the practice. Both ancient and modern knowledge have found institutional grounding. In traditional societies, young individuals glean knowledge from family elders. In today's context, structured education and pedagogical frameworks drive industrial progress. As observed in countries like Japan and Thailand, structured learning and training are essential for preserving heritage [47].

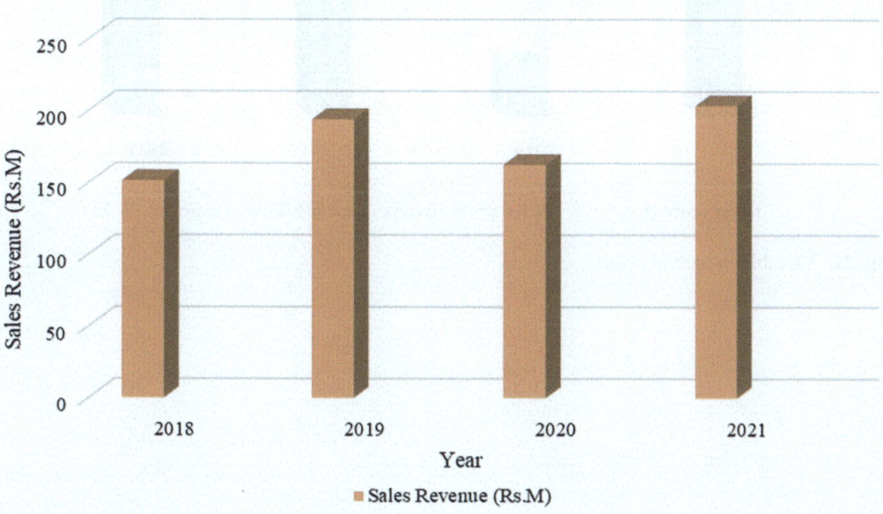

Fig. 19 Southern province sales revenue

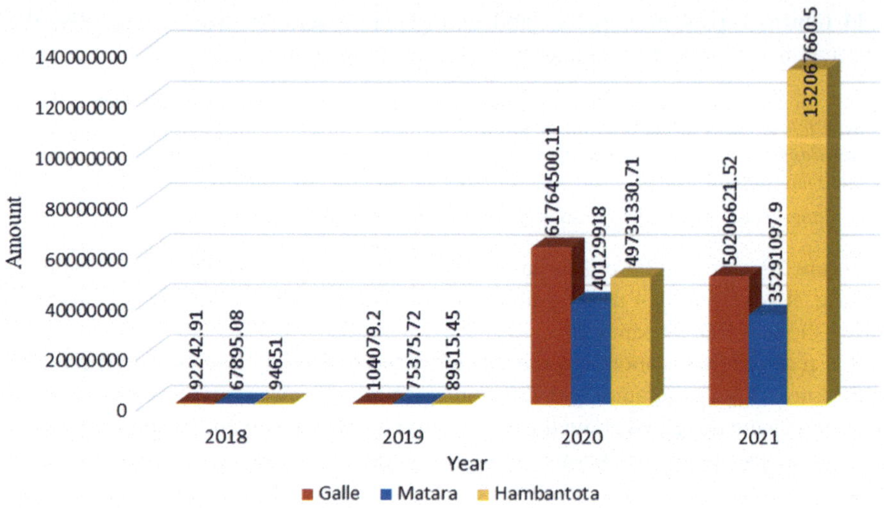

Fig. 20 Variation in production during the last four years

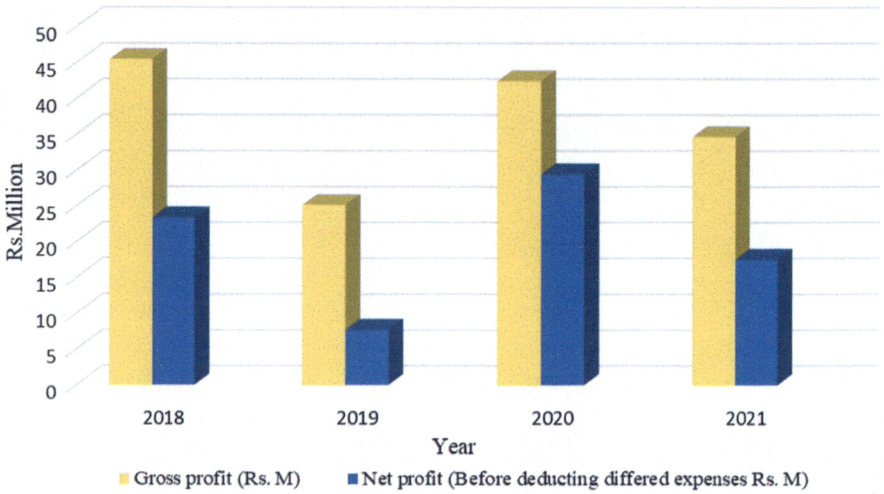

Fig. 21 Variable in annual profit

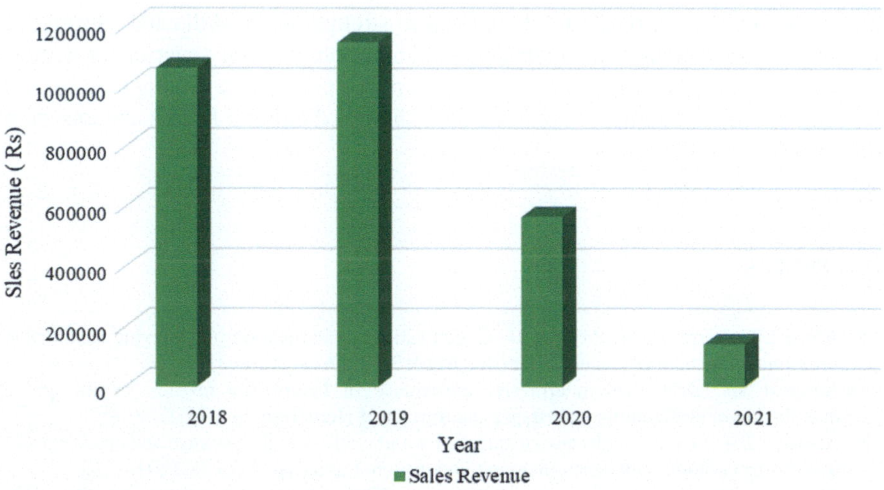

Fig. 22 Online sales revenue of "Ruhunu Ransalu Fashion"

19 Conclusion

Sri Lankan handloom weaving is an outstanding manifestation of ethical craftsmanship and environmental sustainability, appealing to environmentally conscious customers and paying tribute to its wide range of yarns, colors, and patterns. The tenacity of the weavers, along with rising output and market demand, has helped the handwoven textile sector secure its position in the economy while preserving its rich cultural heritage. However, handloom weavers face obstacles like precarious financial situations and the generational transmission of their trade, endangering both their livelihoods and the viability of hand weaving. Fostering partnerships, boosting skill development, holding national competitions, and showcasing their works are beneficial for reviving handloom textile weaving while considering the interconnected cultural, economic, and social conditions.

The study reveals the three characteristics of handloom textile weaving: heritage prevails as a living entity and living skills toward advancing Sustainability Goals 01 and 04. Continuation of design practice, fostering traditional and modern design concepts, and continuation of advancing skills and skill preservation.

In conclusion, the study focused on the survival of handloom textile weaving as a living tradition in Sri Lanka's Southern Province. The study demonstrates that handloom weaving is more than just a historical art form; it is a dynamic and ingrained part of the area's cultural tapestry. This vitality is demonstrated by the weavers' commitment to protecting and promoting their legacy, guaranteeing its ongoing relevance in the modern world. It is evident from this study's findings that the Southern Province handloom weaving community merits support and acknowledgment from the Provincial Council and the Central Government for the enhancement of skills development and learning. The initiatives to promote information exchange, skill improvement, and broader market outreach are demanded to strengthen the skills and learning of the weavers and increase sustainability. Collaboration between government bodies, academic institutions, and cultural groups can play a pivotal role in safeguarding and elevating the art of handloom weaving as a cherished cultural legacy and ensuring its stability.

References

1. Abreu MC, Ferreira FN, Proenca JF, Ceglia (2021) Collaboration in achieving sustainable solutions in the textile industry. J Bus Ind Mark 36(9):1614–1626
2. Agrawal SA (2020) The interpretive experience of handwoven textiles: exploring the intersection of sustainability, heritage, and tourism. J Herit Tour 738–757:738–757
3. Ahern J (2013) Urban landscape sustainability and resilience: the promise and challenges of integrating ecology with urban planning and design. Landscape Ecol 28:1203–1212
4. Ariely D, Gneezy U, Loewenstein G, Mazar N (2009) Large stakes and big mistakes. Rev Econ Stud 76(2):451–469
5. Arvey RD, Campion JE (1982) The employment interview: a summary and review of recent research 1. Pers Psychol 35(2):281–322
6. Banadara A, Miss de Alwis I, Abeysinghe S, Munid GDP, Kalyani WP, Wijesinghe HARN, Chamari Fonseka M, Hapuarachchi U, Pathiraja PANC, Jayawardhane WASR, Madagedara GR, Fernando Mrs WAK, Priyashanika GHRA, Weerasinghe AAS, Nawarathna WPDM, Senani KAML, Nalani M, Priyanka Mr SDC, Bandara Mrs HMRA, Dinithi Madushani R, Jayaweera ML (2022) Industry data book
7. Barbier EB (2020) Greening the post-pandemic recovery in the G20. Environ Resource Econ 685–703
8. Baxter P, Jack S (2015) Qualitative case study methodology: study design and implementation for novice researchers. Qual Rep 13(4):544–559
9. Becker HS (1982) Art worlds. University of California Press
10. Becker HS (1978) Arts and crafts. Am J Sociol 83(4). https://about.jstor.org/terms
11. Bernstein N (1967) The coordination and regulation of movements. Pergamon Press
12. Bourdieu P (1977) Outline of a theory of practice. Cambridge University Press
13. Bourne M, Neely A, Platts K, Mills J (2002) The success and failure of performance measurement initiatives: perceptions of participating managers. Int J Oper Prod Manag 22(11):1288–1310
14. Braun V, Clarke V (2006) Using thematic analysis in psychology. Qual Res Psychol 3(2):77–101. https://doi.org/10.1191/1478088706qp063oa
15. Bril B (2018) Action, movement, and culture: does culture shape movement? Kinesiol Rev 7(1):79–87
16. Bril B (1986) Motor development and cultural attitudes. In: Themes in motor development. Springer, Netherlands
17. Bril B, Zack M, Nkounkou-Hombessa E (1989) Ethnotheories of development and education: a view from different cultures. Eur J Psychol Edu 307–318
18. Brinkmann Kvale S (2018) Doing interviews. SAGE
19. Brown JD (2005) Research methods for applied linguistics. In: Davies A, Elder C (ed) Blackwell
20. Bruckner M, Hosek E, Jussila R, Karst L, Kefalidou A, Kuehner M, Li M, Rambler S, Rauch S, Soltau F, Tagashira M, Zubcevic I (2018) Synthesis of voluntary national reviews 2018
21. Bryman A (2008) Social research methods. Oxford University Press
22. Bull C, Schotter A, Weigelt K (1987) Tournaments and piece rates: an experimental study. J Polit Econ 95(1):1–33

23. Burch S, Shaw A, Robinson J (2014) Triggering transformative change: a development path approach to climate change response in communities. Climate Policy 14(4):467–487
24. Camerer C (2005) Three cheers—psychological, theoretical, empirical—for loss aversion. J Mark Res 42(2):145–153
25. Cameron C (2010) The unnatural history of heritage: what is the future for the past? 218
26. Carey MA (1994) The group effect in focus groups: planning, implementing, and interpreting focus group research. Crit Issues Qual Res Methods 225:41
27. Coomaraswamy A (1913) Art and crafts of India and ceylon. Today and Tomorrow's Printers and Publishers
28. Coomaraswamy AK (1959) Medieval sinhalese art. Published by the National Museum
29. Crabtree BF (1999) Doing qualitative research. SAGE
30. Cross N (1999) Designerly ways of knowing: design discipline versus design science. Des Issues 17(3):49–55
31. Cross N (1982) Designerly ways of knowing, design studies, 1st edn, vol 3
32. Daes EIA (1995) Protection of the heritage of indigenous people: final report of the Special Rapporteur, Erica-Irene Daes, in conformity with Subcommission resolution 1993/44 and decision 1994/105 of the Commission on Human Rights
33. Dahling JJ, O'Malley AL, Chau SL (2012) Effects of feedback motivation on transfer-of-training: multilevel analysis of individual differences. Hum Resour Manag J 22(2):180–195
34. Deci EL, Koestner R, Ryan RM (1999) A meta-analytic review of experiments examining the effects of extrinsic rewards on intrinsic motivation. Psychol Bull 125(6):627–668
35. Del Barrio MJ, Herrero LC, Sanz JÁ (2009) Measuring the efficiency of heritage institutions: a case study of a regional system of museums in Spain. J Cult Heritage 10(2):258–268
36. Denzin NK (2001) The reflexive interview and a performative social science. Qual Res 1(1):23–46
37. DiCicco-Bloom B, Crabtree BF (2006) The qualitative research interview. Med Edu 40(4)
38. Dörnyei Z (2007) Research methods in applied linguistics: quantitative, qualitative, and mixed methodologies. Oxford University Press
39. Edmondson A (2019) The fearless organization: creating psychological safety in the workplace for learning, innovation, and growth. Wiley
40. Endresen K (1999) Sustainable tourism and cultural heritage
41. Fine GA (2003) Crafting authenticity: the validation of identity in self-taught art. Theory Soc 153–180
42. Gandon E, Nonaka T, Sonabend R, Endler J (2020) Assessing the influence of culture on craft skills: a quantitative study with expert nepalese potters. Plos One 15(10)
43. Gneezy U, Meier S, Rey-Biel P (2011). When and why incentives (do not) work to modify behavior. J Econ Perspect 25(4):191–210
44. Graham B (2002) Heritage as knowledge: capital or culture? Urban Stud 39:1003–1017
45. Groth C (2017) Making sense through hands—design and craft practice analysed as embodied cognition. Aalto University
46. Groulx M, Nowak N, Levy K, Booth A (2021) Community needs and interests in University-Community partnerships for sustainable development. Int J Sustain High Educ 22(1):274–290. https://doi.org/10.1108/IJSHE-03-2020-0086
47. Hancock B (2002) An introduction to qualitative research. Trent Focus Group
48. Hardon A, Hodgkin C, Fresle D (2004) How to investigate the use of medicines by consumers. World Health Organization and University of Amsterdam
49. Harper D (1987) Working knowledge: skill and community in a small shop. University of Chicago Press
50. Harrison R (2010) What is heritage? Underst Politics Heritage 5–24
51. Hattie J, Timperley H (2007) The power of feedback. Rev Edu Res 77(1):81–112
52. Holland BA (2005) Scholarship and mission in the 21st-century university: the role of engagement. In: Proceedings of the Australian Universities quality forum, vol 5. AUQA Occasional Publication

53. Holtorf C (2018) Embracing change: how cultural resilience is increased through cultural heritage. World Archaeol 50:639–650
54. Huotilainen M, Rankanen M, Groth C, Seitamaa-Hakkarainen P, Mäkelä M (2018) Why our brains love arts and crafts: Implications of creative practices on psychophysical well-being 11(2):118
55. ILO (2004) Human resources development recommendation
56. ILO (2008) Conclusions on skills for improved productivity, employment, growth, and development. In: International labour conference
57. Ingold T (2001) From the transmission of representations to the education of attention. In: Whitehouse H (ed) The debated mind: evolutionary psychology versus ethnography, pp 113–153
58. Jacobs DM, Michaels CF (2007) Direct learning. Ecol Psychol 19:321–349
59. Johnson R (1993) Introduction. In: Bourdieu P (ed) The field of cultural production. Polity Press
60. Kallio H, Pietilä AM, Johnson M, Kangasniemi M (2016) Systematic methodological review: developing a framework for a qualitative semi-structured interview guide. In: Journal of advanced nursing, vol 72, issue no 12. Blackwell Publishing Ltd., pp 2954–2965. https://doi.org/10.1111/jan.13031
61. Kautish P, Khare A, Sharma R (2020) Values, sustainability consciousness, and intentions for SDG endorsement. Mark Intell Plan 38:921–939
62. Kelly C (2009) Heritage. University of Greenwich
63. Kim EKA et al. (2020). Interpretation of Korean traditional textiles with a focus on sustainability. Sustainability
64. Lam T (2006) The politics of heritage tourism in China: a view from Lijiang. J Sustain Tour 300–316
65. Lave J, Wenger E (1991) Situated learning legitimate peripheral participation. Cambridge University Press
66. Lazear EP (2000) Performance pay and productivity. Am Econ Rev 90(5):1346–1361
67. Loewenstein G, Prelec D (1992) Anomalies in Intertemporal choice: evidence and an Interpretation. Quart J Econ 107(2):573–597
68. Loh P (2016) Community–university collaborations for environmental justice: toward a transformative co-learning model. New solutions. J Environ Occup Health Policy 26(3):412–428
69. Magdalena PS (2018) Cultural heritage in EU policies 11
70. Makhitha KM (2013) An investigation into buyer behaviour of craft retailers in South Africa [PhD thesis]. University of Pretoria
71. Marchand THJ (2008) Muscles, morals, and mind: craft apprenticeship and the formation of person. Br J Educ Stud 56(3):245–271
72. Matos AA (2018) The cultural and social importance of traditional textiles: interpretation and intervention strategies. Int J Heritage Stud 665–682
73. Mazzucato M (2019) The value of everything: making and taking in the global economy. Penguin Books
74. Mills CW (1980) On intellectual craftsmanship 17(2):63–70
75. Moore-Cherry N, Whelan Y (2016) Heritage, memory, and the politics of identity
76. Mortara M, Catalano CE, Bellotti F, Fiucci G, Houry-Panchetti M, Petridis P (2014). Learning cultural heritage by serious games. J Cult Heritage 15(3)
77. Mtawa NN, Wangenge-Ouma G (2016) The scholarship of university-community engagement: interrogating Boyer's model. Int J Educ Dev 49:126–133
78. Murray K (2011) Sustainability in craft and design. Craft+ design enquiry, 3rd edn. ANUE Press
79. Oakley ACJ (1981) Subject women. Pantheon Books
80. Otero J (2022) Heritage conservation future: where we stand, challenges ahead, and a paradigm shift. Glob Challenges 6
81. Pasikowska-Schnass M (2018) Cultural heritage in EU policies. Eur Parliamentary Res Serv 11

82. Patton MQ (1990) Qualitative evaluation and research methods, vol 2. SAGE, p 532
83. Di Pietro L, Guglielmetti Mugion R, Renzi MF (2018) Heritage and identity: technology, values and visitor experiences. J Herit Tour 13(2):97–103
84. Polit DS, Beck CT (2010) Essentials of nursing research. Appraising evidence for nursing practice, 7th edn. Lippincott-Raven Publishers
85. Porter L, Butt A, Rosier J, Perkins T, Crookes L, Barry J (2015) Partnerships of learning for planning education: who is learning what from whom? Beautiful Messiness Learn Partnerships Plann Theory Pract 16(3):409–434
86. Rahaman H (2018) Digital heritage interpretation: a conceptual framework. Dig Creat 29:208–234. https://doi.org/10.1080/14626268.2018.1511602
87. Reed ES, Bril B (1996) The primacy of action in development. Dexterity Dev 1:431–451
88. Renzi LD (2014) Cultural technology district: a model for local and regional development. Curr Issue Tour 17:640–656
89. Renzi LD (2018) Heritage and identity: technology, values and visitor experiences. J Heritage Tour 13(2):97–103
90. Rio Tinto (2011) Why cultural heritage matters. http://www.Riotinto.Com/Documents/ReportsPublications/Rio_Tinto_Cultural_Heritage_Guide
91. Robson C (2011) Real-world research: a resource for users of social research methods in applied settings
92. Rollnick S, Miller WR (1995) What is motivational interviewing? Behav Cogn Psychother 23(4):325–334
93. Rubin DB (2005) Causal inference using potential outcomes: design, modeling, decisions. J Am Stat Assoc 97–103
94. Rubin HJ, Rubin IS (2011) Qualitative interviewing: the art of hearing data. SAGE
95. Ryan RM, Deci EL (2000) Intrinsic and extrinsic motivations: classic definitions and new directions. Contemp Educ Psychol 25(1):54–67
96. Santos MC (2019) The challenge of sustaining traditional crafts: collaborations between artisans and designers. J Text Sci Technol 15–22
97. Seidman I (2012) Interviewing as qualitative research: a guide for researchers in education and the social sciences. Teachers College Press
98. Sennett R (2009) The craftsman. Penguin Books
99. Serry T, Liamputtong P (2013) The in-depth interviewing method in health 39–53
100. Sethi R (2022) Handmade for the 21st century: safeguarding traditional Indian textiles. ISBN: 978-81-89218-84-3
101. Sethi R, Duque Duque C, Vencatachellum (2005) Designers meet Artisans—a practical guide. UNESCO, Craft Revival Trust
102. Shulla K, Voigt BF, Cibian S, Scandone G, Martinez E, Nelkovski F, Salehi P (2021) Effects of COVID-19 on the sustainable development goals (SDGs). Discov Sustain 2(1). https://doi.org/10.1007/s43621-021-00026-x
103. Shute VJ (2008) Focus on formative feedback. Rev Educ Res 78(1):153–189
104. Silverman HA (2008) Heritage, memory and the politics of identity: new perspectives on the cultural landscape. University Press of Florida
105. Smith L (2006) The uses of heritage
106. Smith L, Akagawa N (2008) Intangible heritage
107. Snyder RS (2013) The role of heritage conservation districts in achieving community economic development goals. Int J Herit Stud 19:304–321
108. Stone D, Heen S (2014) Thanks for the feedback: the science and art of receiving feedback well. Penguin
109. Sustainable Development Act (2017) Sri Lanka Sustainable Development Act, No. 19 OF 2017. www.documents.gov.lk
110. Taylor MC (2005) Interviewing. In: Qualitative research in health care. McGraw-Hill Education
111. Tertiary and Vocational Education Commission (1992) The first development plan for tertiary and vocational education in Sri Lanka

112. Thorlindsson T, Halldorsson V, Sigfusdottir ID (2018) The sociological theory of craftsmanship: an empirical test in sport and education. Sociol Res Online 23(1):114–135
113. Timothy DJ, Boyd SW (2003) Heritage tourism. Prentice Hall
114. UNESCO (1989) Medium term plan 1990–1995
115. UNESCO (2009) Cultural heritage and the challenge of sustainability
116. UNESCO (2022) Reimagining our futures together: a new social contract for education
117. UNESCO Convention (2003) Text of the convention for the safeguarding of the intangible cultural heritage. https://ich.unesco.org/en/convention
118. United Nations (2000) United Nations millennium declaration
119. United Nations (2015) Transforming our world: the 2030 agenda for sustainable development
120. United Nations (2023) Why are green skills important for youth? United Nations climate change. https://Unfccc.Int/News/Why-Are-Green-Skills-Important-for-Youth
121. Uzzell DL (1996) Creating place identity through heritage interpretation. Int J Herit Stud 1:219–228
122. Veblen T (1914) The instinct of workmanship and the state of the industrial arts. Palgrave Macmillan
123. Vision (n.d.) A country enrichment : vision 2025. https://www.News.Lk/Images/Pdf/2017/Sep/Vision_2025_English
124. Wals AE, Schwarzin L (2012) Fostering organizational sustainability through dialogic interaction. Learn Organ 19(1):11–27
125. Warren WH (2006) The dynamics of perception and action. Psychol Rev 113–358
126. Waterton EA (2010) Culture, heritage, and representation: perspectives on visuality and the past. Ashgate Publishing, Ltd
127. Watson EW (2015) The Palgrave handbook of contemporary heritage research. Palgrave Macmillan
128. Wijayapala UGS, Alwis AAP, Ranathunga GM, Karunaratne PVM (2022) Evolution of Sri Lankan textile education from ancient times to the 21st century. In: Textile and fashion education internationalization: a promising discipline from South Asia. Springer, Singapore, pp 119–144
129. Winstone NE, Nash RA (2016) The developing engagement with feedback toolkit. J Perspect Appl Acad Pract 4(3):35–42
130. Yin RK (2005) Case study research: design and methods. SAGE, New York
131. Yin RK (2009) Case study research: design and methods. SAGE
132. Zhang HQ (2005) Tourism and hotel development in China: from political to economic success. Haworth Hospitality Press and International Business Press

The Potential of Utilizing Sri Lankan Banana Agro-Waste to Generate Sustainable Textiles

G. L. D. Wickramasinghe, U. S. W. Gunasekera, and A. C. S. I. Mumthas

Abstract This chapter aims to provide available data on banana varieties of Sri Lanka, potential opportunities for extracting fibre from common varieties, their physical, chemical, and morphological characteristics, and possible textile applications and related government interventions. Banana is the most widely grown fruit crop in Sri Lanka and generates massive organic waste. This waste, a rich source of cellulose fibre, can be used to extract sustainable natural fibres as an alternative in the textile industry. A limited number of studies have been conducted on Sri Lankan banana fibres' physio-chemical and morphological characteristics and their applications in textiles. Present literature on banana fibres are country-specific factors; therefore, an overall analysis and their application in the Sri Lankan context will generate new market opportunities and open avenues for further research. The chapter will address SDGs 1 and 9.

Keywords Sri Lankan banana · Banana varieties · Banana fibre · Pseudo-stem fibre · Banana textile

Linkage to SDGs

The chapter addresses SDG 9 by providing opportunities to utilize agro-waste to generate value-added products and SDG 1 by emphasizing the opportunities to generate additional income for rural farmers.

G. L. D. Wickramasinghe · U. S. W. Gunasekera
Department of Textile and Apparel Engineering, University of Moratuwa, Moratuwa, Sri Lanka
e-mail: dharmasri@uom.lk

U. S. W. Gunasekera
e-mail: ujithe@uom.lk

A. C. S. I. Mumthas (✉)
Sri Lanka Institute of Textile and Apparel, Kandawala Estate, No. 02, Gen. Sir John Kotelawela Riad, Ratmalana, Sri Lanka
e-mail: imumthas@gmail.com

G. M. Ranathunga (ed.), *Textiles of Sri Lanka*, SDGs and Textiles,
https://doi.org/10.1007/978-981-97-0569-6_4

Character	M. Accuminata	M. Balbisiana
Pseudostem colour	More or less heavily marked with brown of black blotches	Blotches very slight or absent
Petiolar canal	Margin erect or spreading, with scarious wings below, not clasping pseudostem	Margin inclosed, not winged below, clasping pseudostem
Peduncle	Usually downy or hairy	Glabrous
Pedicels	Short	Long
Ovules	Two regular rows in each loculus	Four irregular rows in each loculus
Bract shoulder	Usually high (ratio < 0.28)	Usually low (ratio > 0.30)
Bract curling	Bracts reflex and roll back after opening	Bracts do not reflex
Bract shape	Lanceolate or narrowly ovate tapering sharply from the shoulder	Broadly ovate, not tapering
Bract apex	Acute	Obtuse
Bract colour	Red, dull purple or yellow outside; pink, dull purple or yellow inside	Distinctive brownish-purple outside; bright crimson inside
Bract colour fading	Inside bract colour usually fades to yellow towards the base	Inside bract colour usually continuous to base
Bract scars	Prominent	Scarcely prominent
Free tepal of male flower	Variably corrugated below tip	Rarely corrugated
Male flower colour	Creamy white	Variably flushed with pink
Stigma colour	Orange or rich yellow	Cream, pale yellow or pale pink

Fig. 1 Fifteen diagnostic characteristics used in taxonomic method. *Source* [35]

1 Introduction

1.1 Scientific Genome of Banana

Southeast Asia is considered the centre of origin and home to more than 70 wild banana species. Musa paradisiaca Linn is the first scientific name given for the plantain cultivar by Karl Linnaeus in 1753, and later, in 1759, Linnaues named the desert banana cultivar Musa sapientum Linn. However, instead of the naming system of Linnaeus, [35] introduced a new naming system originating from Musa Accuminata Colla (A) and Musa balbisiana Colla (B), which are wild parental species of banana endemic to Southeast Asia. Furthermore, [35] introduced a three-tier naming system and taxonomic method using 15 diagnostic characters to segregate the banana varieties into six genome groups.

The 15 characters shown in Fig. 1 were assigned either Score 1 or 5 based on the closeness to the Genomes A and B, respectively. Based on the total score obtained, the genome groups were allocated [35, 38].

Upon careful evaluation of scores for 15 distinct characteristics, bananas are classified into specific genome groups. If the score falls within the range of 15–25, it is grouped under AA/AAA. A score between 26 and 46 leads to placement in the AAB group, while a score of 47–58 corresponds to the AB group. Bananas with a score of 59–63 are placed in the ABB group and 67–69 in the ABBB group. Finally, those with a score between 70 and 75 are classified under BB/BBB group [38]. However, [7] Ekanayake et al. stated that the AAB genome is a complex group with nine specific forms.

1.2 Sri Lankan Banana Varieties

Chandrarathna and Nanayakkara [4] reported 29 Sri Lankan banana varieties.

Currently, the gene bank of the Plant Genetic Resource Centre comprises 22 recorded varieties, alongside two wild species and two unclassified types. Of the 26 varieties, only five belong to the cooking category, while the remaining belong to the dessert category. Detailed classification of Sri Lankan banana varieties based on their genomes is presented in Table 1.

Aside from the primary varieties, numerous genetically modified variations have been engineered to enhance output. A compilation of some of these varieties can be found in Table 2. Sri Lanka's diverse climatic regions accommodate the growth of bananas, with Table 3 outlining suitable varieties for each zone according to their specific growth demands.

Even though most cultivated areas are limited to these three climatic zones, banana can be grown in any area within 1500 m from sea level, with an annual rainfall between 1250–1900 mm and a temperature range of 15–30 °C.

Table 1 Sri Lankan banana varieties and their genetic names

Genome group	Name of the varieties in Sri Lanka
AA	Unel/Unakehel
BB	Eti Kehel
AAB	Embul/Ambul, Suwendel, Muwanthikehel, Rathkehel and Kolikuttu, Puwalu
AAA	Anamalu, Embon (Ambun), Bin Kehel, and Nethrapalam
ABB	Seeni, Alu kehel, Atamuru (Etamuru), Hambanpuwalu, Marathamana, Mondon, and Sambel

Source Ploetz et al. [30], Weerasooriya [40], Liyanage et al. [20], De Costa and Chandima [6], Perera et al. [24], Nazriya et al. [28]

Table 2 New banana varieties

Name of the variety	Main variety	Developed year
Nadee (AAB)	Ambul	1996
Agraa (AAB)	Kolikuttu	2005
Parakum (ABB)	Seeni Kesel	2005
Kandula (AABB)	Tetroploid hybrid banana variety	2002
Pulathisi (ABBB)	Cooking variety	2002
Gannoruwa Seeni Kesel (ABB)	Seeni Kesel	2016
Gannoruwa Rath kesel (AAB)	Rath Kesel	2016
Prasad (ABB)	Alu kesel	2002
Millawa Suwandel (AAB)	Suwandel	2010

Source Mankotte and Department of Agriculture [21]

Table 3 Banana varieties based on climatic zones

Climatic zone	Desert varieties	Cooking varieties
Wet zone	Kolikuttu	Alu Kesel
	Ambul	Mondan
	Seeni Kesel	Pulathisi
	Kandula	Atamuru
		Kandula
Intermediate zone	Kolikuttu	Alu kesel
	Ambul	Mondan
	Seeni	Atamuru
	Anamalu	Kithala
	Ambun	Pulathisi
	Kandula	Kandula
Dry zone	Ambul	Mondan
	Seeni	
	Anamalu	Atamuru
	Ambun	
	Suwadel	Kithala
	Puwalu	
	Bin Kesel	Pulathisi
	Nethrampalam	
	Rath Kesel	Kandula
	Kandula	

Source Mankotte and Department of Agriculture [21]

1.3 History

Bananas have a rich history dating back to 500 BC when they were first mentioned in Sanskrit texts. They were also recorded in historical texts from Hindu, Chinese, Greek, and Roman cultures. It is believed that bananas originate from the jungles of Southeast Asia and the Western Pacific region. The fruit was introduced to Europe by the armies of Alexander the Great in 327 BC, and Arab traders brought it to Africa. Portuguese explorers introduced bananas to the Canary Islands, and Spanish and Portuguese travellers introduced different varieties to various parts of the world. The Caribbean and Latin American countries were introduced to the fruit in 1516. With the advent of refrigerated maritime transportation in the late-fourteenth century, bananas became an internationally traded commodity. Today, over 1000 varieties of bananas are available worldwide, grown in more than 120 countries. India, Uganda, Ecuador, Brazil, Colombia, China, the Philippines, Indonesia, the Democratic Republic of Congo, and Costa Rica are the top ten banana producers worldwide.

Sri Lanka's rich history is intertwined with bananas, which have been cultivated since the Stone Age. Evidence of ati-eta Kesel banana seeds carbonized over time was discovered in the cave sites of the Rathnapura district. Additionally, phytolith evidence was unearthed during an excavation at Fahien Rockshelter in Southwestern Sri Lanka. As early as 341 AD, King Buddhadasa noted the medicinal properties of various parts of the banana tree in the first written reference, Saratha Sangragahaya [15, 31]. Today, bananas are still widely cultivated in Sri Lanka and used for both cooking and desserts.

1.4 Current Market Status

In today's global export trade, bananas are a crop of great significance, ranking as the fifth most important crop and the second most important fruit crop. They are cultivated on a wide scale in developing countries located in tropical and subtropical regions of Asia, Latin America, and Africa. In 2022, the annual export of bananas reached a staggering 19.6 million metric tonnes, with a total production of around 99 million, including domestic usage [9]. However, the accuracy of production reports may be influenced by the lack of statistics on subsistence-level home gardening and small-plot cultivation [2].

Sri Lanka is one of the countries that extensively cultivates bananas, utilizing over 54% of the land for fruit cultivation. The country has allocated nearly 60,000 hectares of land to banana cultivation, producing an annual yield of about 780,000 metric tonnes. Bananas are available throughout the year, and households spend approximately 0.16% of their monthly food expenditure on them. The most popular districts for banana cultivation in Sri Lanka are Kurunegala, Rathnapura, Hambanthota, Monaragala, Ampara, and Jaffna, spanning across the wet, intermediate, and dry zones. However, bananas are also cultivated throughout Sri Lanka as a household crop [32].

1.5 Banana Plant and Its Usage

The banana plant is a highly versatile crop, with all parts of the plant serving a specific purpose. Bananas themselves are a nutritious fruit that can be eaten raw or used in various food products. The flowers of the banana plant are also consumed for their medicinal properties, particularly for diseases such as diabetes, dysentery, bronchitis, and ulcers. In addition, the roots, ashes, seeds, and peels of the banana plant have medicinal uses. The young leaves are used to soothe skin irritation, while the inner stem is consumed for its medicinal value. Banana leaves have several uses, including in food processing, packaging, and fibre extraction. The peel and pseudostem are used for fibre extraction as well, with banana peel fibres being used in the paper industry and pseudo-stem and leaf fibres utilized in textiles, non-woven, and composites. The sap from the pseudo-stem can be consumed as a syrup or applied to relieve stings and bites. Additionally, it is used in the manufacturing of organic fertilizers and as a mordant in the textile industry for dyeing. It is also utilized as a fire-retardant finishing material [3, 14].

However, the pseudo-stem of the banana tree is often considered waste once the fruits are harvested. For every tonne of production, four tonnes of waste are generated, and waste generation per hectare of banana production is 220 tonnes [37]. Incorrect management of this waste can lead to serious environmental issues, including the emission of toxic gases such as CO_2 and the growth of harmful fungi that attack the remaining trees [1, 10]. Moreover, the waste cannot be used for animal feed. Therefore, farmers often see it as a burden to remove this waste from their fields, incurring additional costs.

Banana plants are incredibly versatile, offering a range of fibres that differ in strength, colour, and staple length, all of which can be put to various uses. By utilizing the pseudo-stem of the plant, farmers can create products that generate additional income while reducing waste during production. This waste can then be repurposed as raw materials in other industries, such as textiles and composites.

While the textile industry has traditionally favoured synthetic fibres for their recyclability and the negative environmental impact of natural fibre farming, the industry is now shifting towards more eco-friendly and sustainable practices. Hence, it includes using potential natural waste materials that have a minimal environmental impact as raw materials, making banana agro-waste a perfect source of fibre for the textile industry.

2 Banana Pseudo-Stem Fibre

The textile industry is shifting towards sustainable and environmentally friendly practices, with developed nations spearheading efforts to create greener supply chains. As a result, there is an increasing demand for eco-friendly products. Science and technology are working together to develop innovative processes to help us sustain our relationship with the ecosystem. For instance, banana plantations often dispose of the stems once the fruits have been harvested, leading to tonnes of avoidable waste. However, such stems can now be transformed into a new source of fibre, reducing our reliance on natural and synthetic fibres that require extra energy and chemicals to produce. This new development could benefit rural farmers by providing them with an additional source of income. It could also pave the way for creating a cottage industry, leading to more job opportunities in rural communities. Marketing this eco-friendly product to tourists and exporting to developed nations could be a great way to achieve success. Statistics show that 60,000 hectares of land are currently used for banana cultivation, with about 825 banana plants grown in 1 ha [11]. Additionally, approximately 260 g of fibres can be extracted from a grown banana stem of 1 m in height [22].

2.1 Common Properties of Sri Lankan Banana Varieties

Different parts of the banana tree provide different fibres. This chapter focuses on banana pseudo-stem fibre, closely related to other bast fibres such as jute and hemp. Ten common banana varieties representing the primary genomes in Sri Lanka underwent a fibre property study in which the fibres were manually extracted to observe their common physical, chemical, and morphological properties.

Figure 2 shows the fibre's physical appearance and SEM images at 666 × magnification. At a microscopic level, the fibres in Banana stems have a unique bundle structure like bamboo and ramie fibres. The cloudy appearance is attributed to the binding substance between the fibres in the bundle. SEM images feature the separation of elementary fibres and damaged surfaces resulting from manual extraction.

In Sri Lanka, different varieties of bananas exhibit varying fibre properties, with even the same variety showing significant differences. Typical fibre length varies from 0.5 to 1.5 m based on the height of the stem. Rathkehel has the highest linear density among the commonly found varieties, while Atamuru has the lowest. The banana fibre's linear density ranges from 12 to 32 tex on average. Atamuru has the strongest fibres, while Seeni Kesel has the weakest, with single fibre strength ranging from 1 to 6 Newtons. The tenacity of the fibres varies from 7 to 40 cN/tex, with Ambun having the highest and Suwandel having the lowest. Atamuru has the highest level of elongation, while Ambun has the lowest. The average elongation is between 0.5 and 3%. The moisture content of the fibres varies from 9 to 11%. Rath Kesel has the largest diameter, while Ambun has the smallest, with diameter variation ranging from 100 to 350 μm.

To gain insight into the physical behaviour of macromolecular systems, it is important to analyze molecular orientation. Fourier transform infrared spectroscopy (FTIR) is an essential tool for examining molecular structures on a larger scale. FTIR investigations primarily focus on crystal structure analysis, which is crucial for determining natural fibres' crystallinity index (CI). These fibres have two distinct regions—amorphous and crystalline—and the CI determines the relative amount of crystalline in the cellulose structure. By analyzing the absorbance ratio A1420/A893, FTIR spectroscopy can be used to calculate empirical CI [25]. Furthermore, FTIR analysis can detect the presence or absence of molecular bonds.

The Bruker/ALPHA FTIR spectrometer, equipped with an attenuated total reflection (ATR) attachment, was utilized to examine macromolecular behaviour. Fibre samples were analyzed according to ASTM E168-06, with the samples being placed directly under the crystal and ATR attachment (Fig. 3).

The analysis of banana fibres using FTIR spectroscopy shown in Fig. 3 revealed the presence of lignin through the intensity of the bands and the asymmetric stretching of the C=C double bond within peaks around 1605 and 1640 cm^{-1}. Atamuru and Aanamalu displayed a higher band intensity, indicating the highest lignin content. On the other hand, Puwalu and Ambun exhibited a lower band intensity, which is a clear indication of lower lignin content.

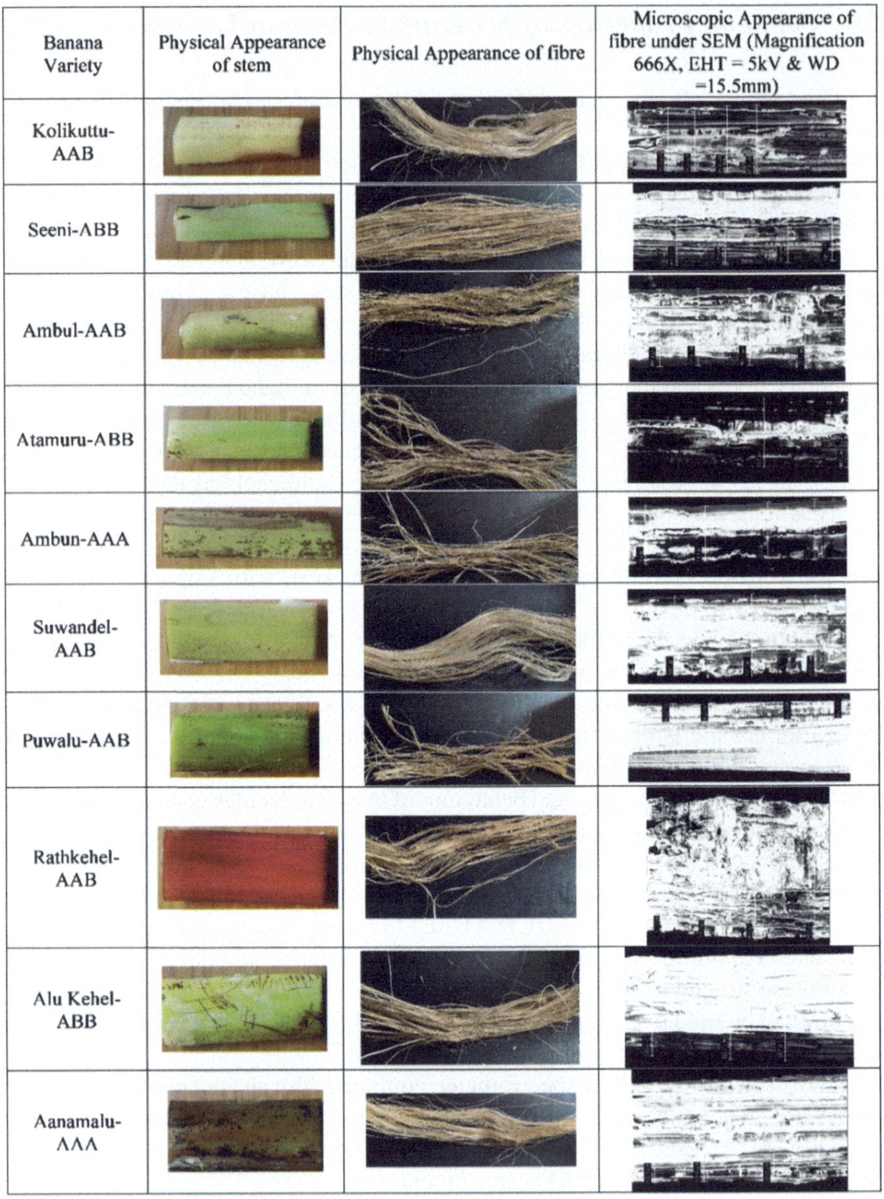

Fig. 2 Physical appearance of selected banana stems and fibres

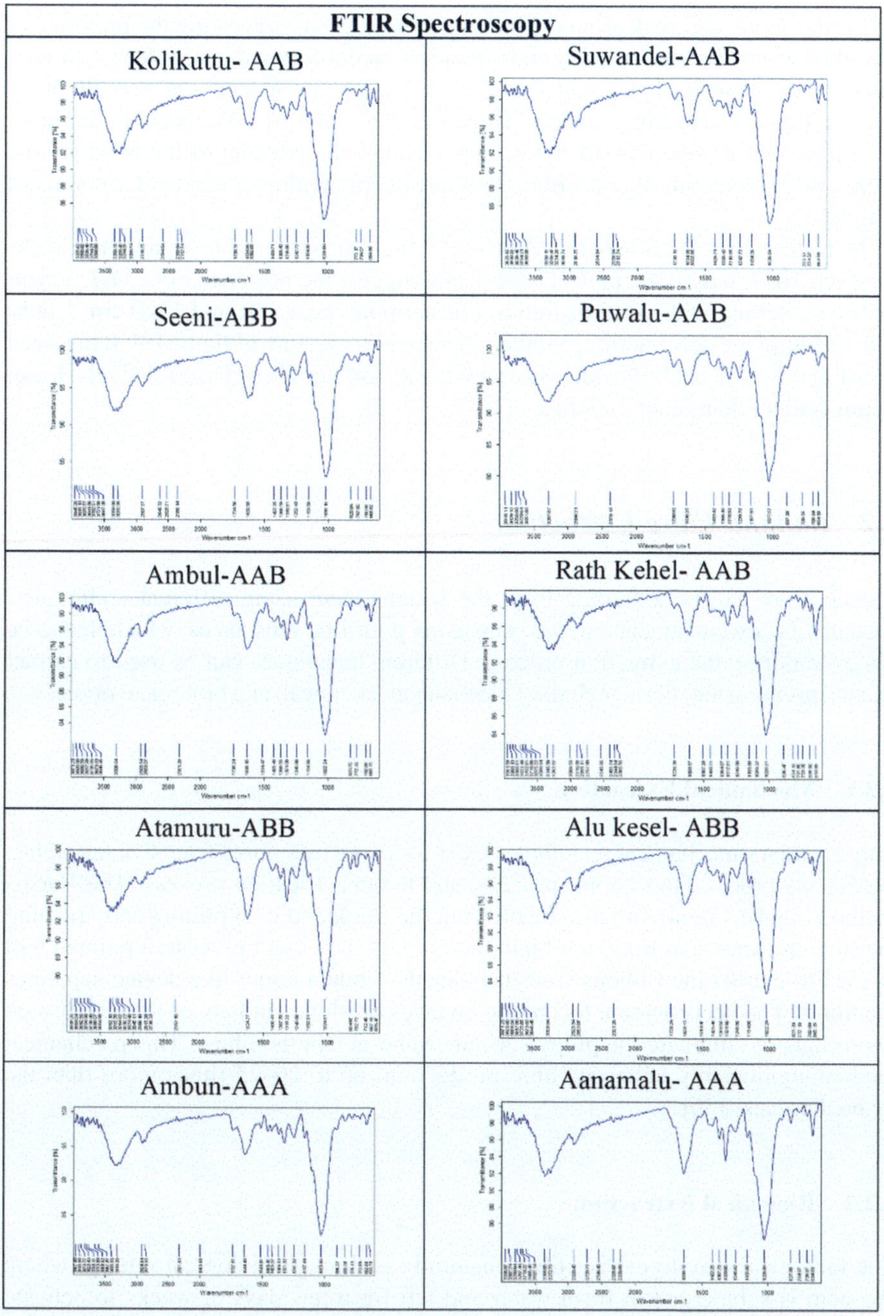

Fig. 3 FTIR spectroscopy of fibres extracted from different banana varieties

Furthermore, the peak around 3400 cm^{-1} broadened, suggesting the presence of absorbed water aliphatic primary and secondary alcohols found in cellulose, hemicellulose, and lignin [16, 17]. Jagadeesh et al. [13] noted that an increased breadth of the hydroxyl group stretching band indicates a higher cellulose percentage in the fibre. It is important to note that all fibres have a similar broadening of the band around 3400 cm^{-1}, indicating that all fibre structures contain almost identical amounts of cellulose.

In the fingerprint region below 1500 cm^{-1}, there are discernible asymmetric peaks between 1020 and 1028 cm^{-1}. These peaks suggest the presence of C–O–C asymmetric stretching, indicative of lignin. Furthermore, peaks around 1320 cm^{-1} indicate OH in-plane deformation, another characteristic feature of lignin [8]. It has been noted that Rath Kehel, Puwalu, Suwandel, and Ambun fibres have relatively lower lignin content than other varieties.

2.2 *Banana Fibre Extraction*

Banana fibre can be extracted from the banana stem using different techniques. Banana fibres were attached to the stem using gum-like substances, which had to be removed during the extraction process. Different techniques can be used to extract banana pseudo-stem fibre, including mechanical, chemical, and biological processes.

2.2.1 Mechanical Extraction

Three primary mechanical techniques exist for extracting fibres from banana stems: the Bacnis process, the Leonit process, and the decortication process. The Bacnis method involves removing the sheath from the trunk and then pulling and crushing the ribbons (known as tuxy) to obtain the fibre. In the Leonit process, a pointed tool is used to extract the ribbons from the sheath, while a comb-like device separates the fibres. The decortication technique involves crushing cut banana stems between two rollers to eliminate the pulpy substance and obtain the fibres. This mechanical method significantly enhances fibre production, up to 20–25 times more than the manual process [36].

2.2.2 Biological Extraction

The biological way to extract fibres naturally is through biological retting, where the stem is submerged in fresh water and left for a few days or weeks to activate microorganisms [30]. The enzyme selection should depend on the substrate's type, composition, size, and lignin content for enzyme-based extraction. Cellulases and pectinases are common enzymes used, with cellulases increasing smoothness by removing fibrils from the surface, while pectinases remove inter-lamellar pectin, a natural adhesive substance between fibres. To extract and process fibres from banana and plantain stems, a combination of Bacillus cereus, Bacillus licheniformis, Bacillus subtilis, and Bacillus polymyxa bacteria can be utilized through the controlled system tank retting (CSTR) method [18].

2.2.3 Chemical Extraction

Various research studies have delved into various chemical extraction methods, including the alkali treatment. This approach facilitates the process of fibrillation and effectively disintegrates composite fibre bundles into finer, higher-quality fibres. Using NaOH in this process reduces the coarseness of the fibres, resulting in exceptional quality. Other substances that can be employed for chemical extraction include sulfuric acid, hydrogen peroxide, protease, and sodium citrate [33].

2.2.4 Banana Fibre Properties with Different Extraction Methods

In the past, extracting banana fibre was a laborious process that required extensive manual work. The skill and experience of labourers were crucial in determining the quality of the fibre. Today, three methods are available for extracting banana fibres: mechanical, chemical, and biological. However, mechanical extraction cannot remove the natural binding material found between fibre bundles [29]. Chemical extraction can lead to environmental pollution. Biological extraction is an eco-friendly option that yields more fibres but involves a lengthy natural retting process. After extraction, degumming is necessary to utilize the fibres fully. This process removes the heavily coated non-cellulosic natural binding material from the cellulosic part of the fibre, allowing for optimal use [12].

The extraction of five common varieties was tested using mechanical decortication, pectinase enzyme, and alkali treatment. The results, which are provided below, are conclusive.

Based on the findings, the biological extraction method yields the strongest fibre across all varieties. Chauhan and Sharma's [5] research suggests that enzymes in biological extraction boost fibre strength by breaking down the lignin component, which acts as a binder. In contrast, mechanical extraction can weaken the fibre by damaging the cellulose component. SEM analysis shown in Fig. 4 also reveals that biologically extracted fibres have intact elementary fibres with minimal binding material remaining on the surface, while mechanically extracted fibres have damaged elementary fibres. Furthermore, the results are supported by the empirical CI calculated using FTIR results. A lower CI indicates a crystallinity content loss, as [25] noted. Furthermore, Robles et al. [33] explain that the mercerization process entails partial destruction of intermolecular bonds caused by the penetration of NaOH into the swollen amorphous regions of the fibre where cellulose molecules are held by

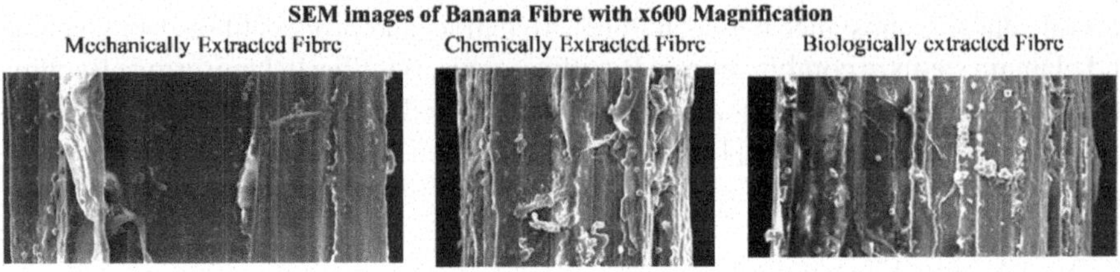

Fig. 4 SEM analysis of fibres with different extraction methods

hydrogen bonding, leading to the formation of sodium cellulose and sodium alcoholates. This process removes a large part of residual hemicellulose and reduces chain lengths due to the rupture of some C–C bonds.

Table 4 illustrate the physical properties of the extracted fibres. The diameter of fibres extracted using mechanical methods is the largest compared to other extraction methods. Fibres extracted through biological methods have a 10% smaller diameter compared to mechanically extracted fibres. Chemical extraction reduces the diameter by 30% compared to mechanically extracted fibres. According to research by Mukhopadhyay et al. [23], banana fibres have a cross-section of several smaller fibres bound together with lignin and hemicellulose structures. These elementary fibres can range from ten to a hundred, and each elementary fibre is comparable to cotton fibrils. SEM analysis in Fig. 4 shows that removing binder material allows for more compact packing of elementary fibres, thereby reducing the overall diameter of the bundle of elementary fibres.

Furthermore, the analysis shows visible empty spaces between elementary fibres, confirming the removal of binder materials, suggesting that deep penetration of

Table 4 Properties of banana fibres with different extraction methods

Property	Fibre extraction method	Ambul	Atamuru	Ambun	Puwalu	Kolikuttu
Single fibre strength (N)	Mechanical	1.431	3.645	2.577	2.769	1.546
	Chemical	1.606	3.428	1.936	2.132	1.615
	Biological	2.155	4.232	3.299	3.794	2.410
Single fibre diameter (μm)	Mechanical	357.958	300.520	140.235	184.055	220.125
	Chemical	141.335	127.500	80.595	105.345	97.820
	Biological	180.985	178.425	101.455	177.750	149.425
Single fibre elongation (%)	Mechanical	2.822	4.272	2.712	4.214	2.268
	Chemical	3.180	2.860	6.502	3.538	2.552
	Biological	2.998	4.128	2.556	5.222	4.274
Tenacity (cN/Tex)	Mechanical	12.123	37.086	33.604	39.480	23.750
	Chemical	16.926	42.942	27.559	36.857	28.926
	Biological	21.739	49.612	45.504	57.810	39.259

enzyme or NaOH into the fibre bundle leads to the breakdown of binder material and separation of elementary fibres from the bundle, resulting in a reduced diameter. Therefore, the remaining binder material in the mechanically extracted fibres occupies some space in the fibre bundle, leading to a larger diameter. Biological and chemical extraction removes the binding material, reducing the space utilized for binders and reducing the diameter.

While the method of extracting fibre does not have a clear correlation with its extensibility, changes in the molecular chains can cause variations in extension. According to Kulkarni et al. [19], even minor alterations in diameter can lead to slight differences in fibre extensibility, but significant variances exist between different varieties. Patel and Patil [27] researched cotton fibre extensibility using various cotton varieties and discovered that extensibility varied between 5 and 13%. Fibre extensibility is also influenced by linear density and moisture content. Furthermore, Warwicker and Hallam [39] noted that treating cotton fibre with alkali can modify its extensibility. As a result, the differences in the extensibility of banana fibre may be due to other related factors and variations in varieties.

The biological extraction method demonstrates the most excellent tenacity, whereas the tenacity of the other two methods varies depending on the variety type. The FTIR analysis (Fig. 5) indicates that fibres extracted via the biological method have a higher CI, which confirms an increase in crystalline regions. This increase contributes to the higher tenacity levels achieved through biological extraction. Furthermore, biological and chemical extraction techniques have been shown to reduce the diameter and increase the strength values of the fibres, leading to higher tenacity levels. Conversely, SEM images of mechanically extracted fibres at various magnifications reveal unclear, cloudy surfaces due to entangled and broken single fibres due to decortication. This damage to the fibres from mechanical extraction is due to their lower tenacity despite the higher binder content, which can contribute to strength. However, binders also occupy space within the fibre, resulting in larger diameters and lower tenacity values.

When fibres are biologically treated and analyzed using scanning electron microscopes (SEM), the resulting images reveal a similar separation to those fibres undergoing chemical extraction. The SEM images provide a visual representation of the removal of binder material from the surface between fibres. In contrast, some binder material may remain on the outer surface of the fibre. Researchers have suggested that this partial removal of non-cellulosic binder material can allow for easier rearrangement of fibrils in the direction of the fibre axis, leading to improved load sharing and increased tenacity. SEM analysis confirms that binder material can be effectively removed without damaging the fibrils, resulting in a smaller diameter and greater tenacity.

Based on the FTIR spectroscopy results obtained from the diagnostic region, the structure appears to contain a C=C double bond and a C_{sp3}-H bond, with sp^3 hybridized carbon above 1500 cm^{-1}. The lignin presence in mechanically extracted fibre is distinguishable by the high band intensity and asymmetric stretching of the C=C double bond, visible in the peaks at 1605 cm^{-1} and 1637 cm^{-1}. Conversely, the band intensity decreases in the 1605–1637 cm^{-1} range for biologically (enzyme)

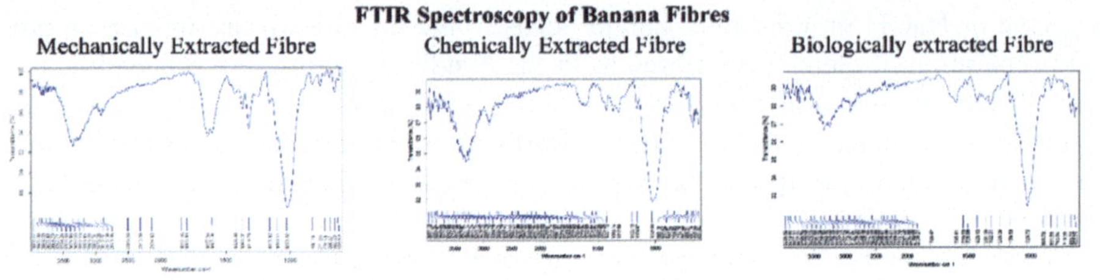

Fig. 5 FTIR spectroscopy of banana fibres extracted from different extraction methods

extracted fibre and becomes insignificant in chemically extracted fibre [8], indicating that lignin reduction occurs through biological and chemical extraction methods.

In addition, the broad peak at around 3400 cm^{-1} indicates the presence of absorbed water aliphatic primary and secondary alcohols found in cellulose, hemicellulose, and lignin [16, 17]. The width of the band increases while the intensity decreases in the order of mechanical, chemical, and biological extraction due to the removal of cementing substances in the inter-fibre space, reducing the total content and opening the avenue to cross-linking. Jagadeesh et al. [13] further suggest that the increased breadth of the hydroxyl group-stretching band implies a higher percentage of cellulose in the fibre, which leads to higher strength.

In the region below 1500 cm^{-1} that pertains to fingerprints, two peaks exist at approximately 1024 and 1318 cm^{-1}. These peaks indicate the C–O–C asymmetric stretching and OH in-plane deformation. The lower intensity of the band around 1318 cm^{-1} in biological extraction suggests a decrease in lignin and an increase in crystallinity. Similarly, the reduced band intensity at about 1024 cm^{-1} in chemical extraction explains the decrease in carbohydrate functional groups. Therefore, FTIR analysis corroborates the outcomes obtained through physical tests.

Based on the ratio provided, it can be inferred that the CI of mechanically, chemically, and biologically extracted fibres was 1.5, 1.1, and 1.6, respectively. The higher CI value in biological extraction is the reason behind its greater tenacity.

This study provides fascinating insights into fibre behaviour with distinct extraction methods. The enzymatic biological extraction resulted in higher values of tenacity and strength, while chemical extraction produced fibres with the smallest diameter values. Biologically extracted fibres exhibited a 40% increase in fibre strength compared to mechanically extracted fibres. Different varieties revealed varying extensibility behaviours, a common trait in natural fibres.

2.3 *Banana Fibre Further Treatments*

Various treatments have been explored in the literature to enhance the characteristics of banana fibres. The bundle structure needs to be broken down to achieve the maximum possible separation, targeting the achievable level of elementary fibre. Numerous supplementary treatments underwent testing to improve fibre characteristics such as texture, flexibility, and fineness.

Mechanically extracted banana fibres were subjected to different chemical treatments as stated below.

Treatment 1: banana fibre treated with 1% pectinase enzyme at 60 °C.

Treatment 2: banana fibre treated with 1% caustic at 90 °C.

Treatment 3: banana fibre treated with 1% glycerin at 90 °C.

Treatment 4: banana fibre treated with 1% hydrogen peroxide at 90 °C.

Treatment 5: banana fibre treated with 1% combined scouring bath having NaOH and H_2O_2 at 90 °C.

Treatment 6: banana fibre treated with 1% sodium bi-sulphide at 90 °C.

SEM results shown in Fig. 6 indicate that all treatments, except the glycerin treatment, successfully removed the gummy substance. The glycerin-treated fibre retained significant gummy residue on its surface. Conversely, the peroxide-treated fibre had damaged elementary fibres still attached to its surface. At the same time, all other treatments achieved a clean surface, indicating that removing the gummy substance was successful.

The physical properties of the fibres after different treatments were tabulated in Table 5. Peroxide treatment has the lowest average fibre diameter, while caustic treatment yields the highest. The average diameter ranges from 45 to 90 μm. The peroxide treatment effectively breaks down individual fibres and separates them from the bundle, reducing the overall diameter. However, the bundle structure remains intact in caustic, leading to a higher diameter value despite removing the gummy substance.

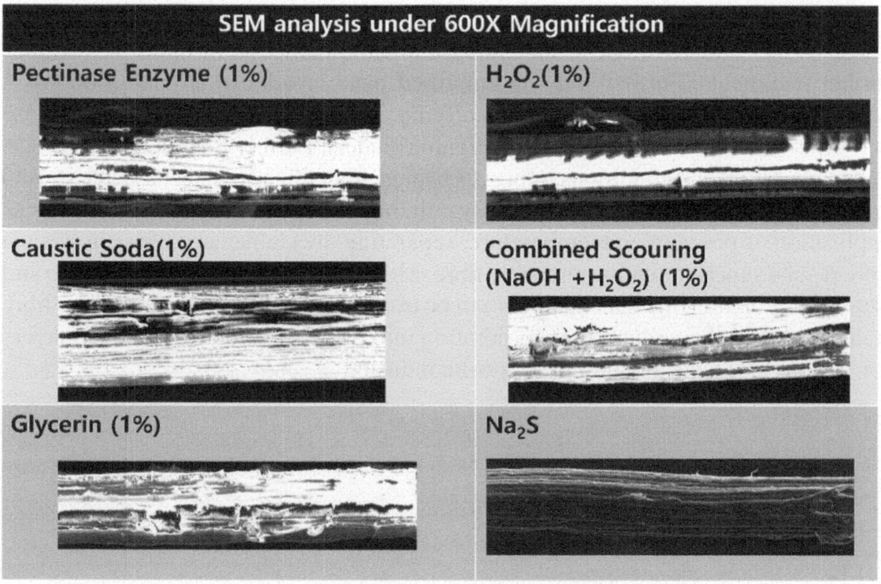

Fig. 6 SEM analysis of fibres with different treatments

Table 5 Fibre properties with different treatments

Treatment	Average fibre diameter (μm)	Average single fibre strength (N)	Average fibre elongation (%)	Average tenacity (cN/Tex)
1	64.75	3.58	0.85	22.95
2	85.63	2.06	4.72	12.00
3	60.10	2.17	3.29	27.15
4	49.23	3.43	3.17	27.96
5	62.27	1.98	2.32	22.10
6	53.31	1.78	2.52	24.34

Fibre that has been treated with enzymes demonstrates the greatest strength, as it prevents any harm to the elementary fibres while eliminating only the gummy substance. On the other hand, the sample treated with sodium bi-sulphide demonstrates the lowest strength, indicating damage to the hemicellulose component while eliminating the gummy substance. The banana fibre sample that underwent caustic treatment exhibited the greatest elongation, in contrast to the sample that underwent enzyme treatment with the lowest. According to Warwicker and Hallam [39], modifying the extensibility of cotton fibre is possible through alkali treatment, and similar results were observed in banana fibre. Peroxide-treated fibre has resulted in the highest tenacity and caustic-treated fibre has resulted in lowest. Robles et al. [33] explained that caustic treatment partially destroys intermolecular bond in cotton fibre. Similar phenomena can be predicted leading to the reduction of fibre strength and tenacity.

The FTIR spectroscopy illustrated in Fig. 7 revealed minimal peaks at approximately 1605 and 1640 cm^{-1}, indicating a notable reduction in lignin levels after further treatments. Furthermore, an amplified band breadth of around 3400 cm^{-1} was observed, with lower intensity, signifying an increased percentage of cellulose and the removal of hemicellulose and lignin (binder). Further treatment with modified temperature and conditions improves banana fibre fineness and characteristics to apply it in general textile applications. By utilizing biochemical treatment and precise application of pressure and temperature, separating the elementary fibres in banana fibre results superior quality spinnable fibre. Moreover, with careful temperature and pressure regulation, the banana fibres can be transformed into a powdered short fibre state, rendering them ideal for incorporation into the production of absorbent layers for sanitary products and in the composite industry.

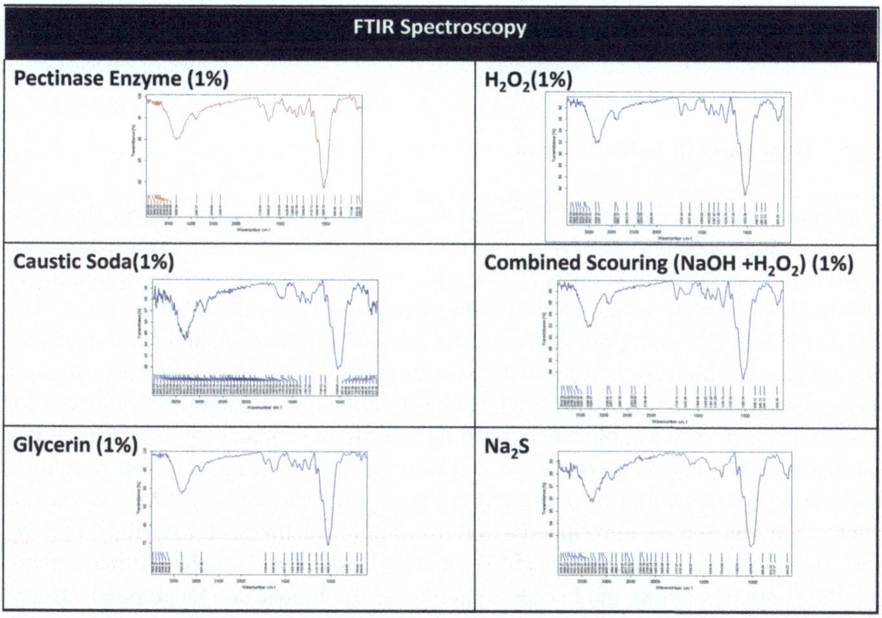

Fig. 7 FTIR spectroscopy of fibres with different treatments

2.4 Applications and Government Interventions

2.4.1 Applications

The banana agro-industry often considers the banana pseudo-stem as waste to be disposed of by the farmers, incurring additional burdens. However, by utilizing this waste as a potential fibre source, the burdens can be reduced, and additional income can be generated through the circularity concept. The entire stem can be used without creating any waste. The outer layers of the stem can be used to extract fibres, and the remaining chopped fibre can be used to create composite or pulp paper. The sap can be used as organic fertilizer and a mordant for natural dye development. The inner core of the stem can be used to create cups or plates using a die arrangement. Banana fibre production can be implemented as a cottage-based industry in Sri Lanka, where rural farmers can use simple decorticator machines to convert waste into valuable products, creating new career opportunities.

The extracted fibre can be used in various industries, from code manufacturing to handcrafted goods to apparel products. Banana fibre has unique characteristics that make it ideal for different applications. It is stronger than other natural bast fibres, seawater-resistant, and naturally buoyant, making it useful for marine applications. Additionally, its absorbency and quick-drying properties make it suitable for absorbent-related applications. After treatments and preparations, it can even be used as a raw material for textile applications. Salit [34] reported that banana fibre can withstand temperatures of up to 200 °C and has inherent anti-bacterial properties with a lifespan of up to three years.

2.4.2 Government Interventions

To increase the contribution of SMEs to the overall economy, the World Bank has recommended measures to alleviate the current economic difficulties Sri Lanka faces. As a result, the government takes a leadership role in cultivating university–industry partnerships to create value-added products through research and innovation. The National Science Foundation, Sri Lanka Inventors Commission, and National Innovation Agency, which operate under the Ministry of Science, offer guidance and assistance in commercializing research. The government has also laid the groundwork for interested investors to establish new startups by utilizing banana agro-waste to create sustainable value-added products. In 2023, the government has devoted roughly 2 billion rupees to developing small and medium-sized enterprises (SMEs). Various loan schemes have been introduced to aid in capital investments for startups. Among them, the 'Small Scale Industries (SSI) credit scheme' was launched in partnership with the Bank of Ceylon, the People's Bank, and the Industrial Development Board (IDB) in 1978. Another credit line, the SMI Loan Scheme, was also established in 1979 to offer finance to SMIs at a reduced interest rate. Currently, numerous other credit schemes are available for SMEs, such as SMILE, SMAP, E-Friends, SPREAP, SMERDP, KfW, Sahanya, Susahana, and Saubhagya [26].

Further the World Bank recommendations to enhance SMEs' contribution to the overall economy are crucial to easing the current economic hardships of Sri Lanka. Therefore, the government also plays a leadership role in fostering university–industry linkages to develop value-added products through research and innovations. Organizations working under the Ministry of Science, such as the National Science Foundation, Sri Lanka Inventors Commission and National Innovation Agency, play significant roles in providing guidelines and support to commercialize research. The government has laid a foundation for interested investors to create new startups through the sustainable utilization of banana agro-waste to create value-added products.

Acknowledgements This work was supported by the Sri Lanka Institute of Textile and Apparel, Ministry of Industries, Sri Lanka, and the research was carried out under the supervision of the academic staff of the University of Moratuwa.

References

1. Ahmad T, Danish M (2018) Prospects of banana waste utilization in wastewater treatment: a review. J Environ Manage 206:330–348. https://doi.org/10.1016/j.jenvman.2017.10.061

2. Anyasi TA, Jideani AIO, Mchau GRA (2013) Functional properties and postharvest utilization of commercial and noncommercial banana cultivars. Compr Rev Food Sci Food Saf 12(5):509–522. https://doi.org/10.1111/1541-4337.12025
3. Balda S, Sharma A, Capalash N, Sharma P (2021) Banana fibre: a natural and sustainable bioresource for eco-friendly applications. Clean Technol Environ Policy 23(5):1389–1401. https://doi.org/10.1007/s10098-021-02041-y
4. Chandrarathna MF, Nanayakkara KDSS (1951) Cultivated varieties of bananas in Ceylon. Trop Agric 107–70–91(2):024422568
5. Chauhan S, Sharma AK (2014) Enzyme treatment in improving the quality of pseudo stem fiber of banana plant to use this bioresource for making handmade paper. Int J Fiber Text Res 4(3):57–61
6. De Costa D, Chandima A (2014) Effect of exogenous pH on development and growth of Colletotrichum musae and development of anthracnose in different banana cultivars in Sri Lanka. J Natl Sci Found 42(3):229. https://doi.org/10.4038/jnsfsr.v42i3.7396
7. Ekanayake EMDSN, Samarasinghe WLG, Randeni RAVK (2001) An unrecorded banana cultivar from Sri Lanka. Ann Sri Lanka Dept Agric 3:37–42
8. Fan M, Dai D, Huang B (2012) Fourier transform infrared spectroscopy for natural fibres. Fourier Transf Mater Anal 3:45–68
9. Food and Agriculture Organization of the United Nations (2022) Banana market review—2022. https://www.fao.org/3/cc6952en/cc6952en.pdf
10. Hussain I, Tarar OM (2014) Pulp and paper making by using waste banana stem. J Mod Sci Technol 2(2):36–40
11. Intercropping Banana in Coconut Lands (nd) In www.cri.gov.lk. Coconut Reasearch Institute of Sri Lanka. Retrieved July 3, 2023, from https://cri.gov.lk/wp-content/uploads/2021/10/c5.pdf
12. Jacob N, Prema P (2008) Novel process for the simultaneous extraction and degumming of banana fibers under solid-state cultivation. Braz J Microbiol 39:115–121
13. Jagadeesh D, Venkatachalam R, Nallakumarasamy G (2015) Characterisation of banana fiber-a review. J Environ Nanotechnol 4(2):23–26
14. Jideani AIO, Anyasi TA (2020) Banana nutrition—function and processing kinetics. IntechOpen eBooks. https://doi.org/10.5772/intechopen.76736
15. Kalyani K (2021) Bananas of Sri Lanka. Lankan Isle. https://lankanisle.lk/index.php/2021/09/20/bananas-of-sri-lanka/
16. Khan MA, Ali KI, Basu SC (1993) IR studies of wood plastic composites. J Appl Polym Sci 49(9):1547–1551
17. Kolboe S, Ellefsen O (1962) Infrared investigations of lignin. A discussion of some recent results. Tappi 45(2):163–166
18. Kudagamage C, Chandrasiri GWJ, Razmy AMZ (2002) Analysis of long-term trends in banana sector of Sri Lanka. Ann Sri Lanka Dep Agric 4:21–32
19. Kulkarni AG, Satyanarayana KG, Rohatgi PK, Vijayan K (1983) Mechanical properties of banana fibres (Musa sepientum). J Mater Sci 18:2290–2296
20. Liyanage ASU, Manawaprema MMC, Mendis MH (1998) Differentiation of A & B genome of banana and plantain (Musa spp.) by esterase enzyme. J Nat Sci Found Sri Lanka 26(2):125–131. https://doi.org/10.4038/jnsfsr.v26i2.3560
21. Mankotte KN, Department of Agriculture (2022) Banana cultivation. Department of Agriculture. https://drive.google.com/file/d/1mgyiBW7Qj1ZjpOHQioto5RidhTUzZBl9/view
22. Manoji A, National Craft Assembly (2008) Banana fibre. National Craft Assembly, Rathnapura
23. Mukhopadhyay S, Fangueiro R, Arpac Y, Şentürk Ü (2008) Banana fibers–variability and fracture behaviour. J Eng Fibers Fabr 3(2):155892500800300200
24. Nazriya NF, De Costa DM, Azhaar AS (2007) Genomic variations of colletotrichum musae morphotypes infecting banana varieties of Sri Lanka. In: Nandalal KDW, Sivayoganathan C, Thevanesam V, Jayaratne DK, Sitheeque MAM, Dewasurendra SD, Sivakanesan R, Perera ADLC, Ariyaratne HBS (eds) Part I- agricultural, biological and medical sciences, vol 12. The University of Peradeniya

25. Nelson ML, O'Connor RT (1964) Relation of certain infrared bands to cellulose crystallinity and crystal lattice type. Part II. A new infrared ratio for estimation of crystallinity in celluloses I and II. J Appl Polym Sci 8(3):1325–1341
26. Nishantha B (2018) Small and medium enterprise sector (SMEs) in Sri Lanka: significance, opportunities and challenges. Certified Manag Accountant 5(4):29–36
27. Patel GS, Patil NB (1989) Cotton fibre breaking tenacity and extension as a function of linear density under standard atmospheric and wet conditions
28. Perera ODAN, Basnayake BMKNK, Karunaratne AM (1999) Physicochemical characteristics, popularity and susceptibility to anthracnose of some local banana cultivars. J Natl Sci Found 27(2):119–130. https://doi.org/10.4038/jnsfsr.v27i2.2985
29. Pitimaneeyakul, U. (2009). Banana fiber: Environmental friendly fabric. In: Proceedings of the environmental engineering association of Thailand, vol 5
30. Ploetz RC, Kelper AK, Daniells J, Nelson SC (2007) Banana and plantain—an overview with emphasis on Pacific island cultivars. In www.traditionaltree.org. Permanent Agriculture Resources. Retrieved July 14, 2016, from https://www.researchgate.net/publication/228645617_Banana_and_plantain_-_an_overview_with_emphasis_on_the_Pacific_Island_cultivars
31. Premathilake R (2018) Sri Lanka's earliest wild Musa bananas? Glob J Archaeol Anthropol 3(2). https://doi.org/10.19080/gjaa.2018.03.555608
32. Ranathilaka MB, Lashmi N, Atukorala W (2019) Production and marketing of banana: estimating the profitability using Walawa region in Sri Lanka. Jbfem 2(1):23–32
33. Robles E, Urruzola I, Labidi J, Serrano L (2015) Surface-modified nano-cellulose as reinforcement in poly (lactic acid) to conform new composites. Ind Crops Prod 71:44–53
34. Salit MS (2014) Tropical natural fibres and their properties. Trop Nat Fibre Compos Prop Manuf Appl 15–38
35. Simmonds NW, Shepherd K (1955) The taxonomy and origins of the cultivated bananas. J Linn Soc London, Bot 55(359):302–312. https://doi.org/10.1111/j.1095-8339.1955.tb00015.x
36. Suhaib AS, Awate NP (2016) A review paper on design and development of banana fiber. Int J Eng Sci Res 5(3)
37. Taib RM, Abdullah N, Sulaiman F, Miskam MA (2014) Characterization of banana (Musa spp.) pseudo-stem and fruit-bunch-stem as a potential renewable energy resource. Int J Biol Vet Agric Food Eng 8(8):712–716. https://www.researchgate.net/publication/265051818_Characterization_of_Banana_Musa_spp_Pseudo-Stem_and_Fruit-Bunch-Stem_as_a_Potential_Renewable_Energy_Resource#fullTextFileContent
38. Valmayor RV, Jamaluddin SH, Silayoi B, Kusumo S, Danh LD, Pascua OC, Espino RRC (1999) Banana cultivar names and synonyms in southeast Asia [Review of banana cultivar names and synonyms in Southeast Asia]. In Molina AB, Roa VN (eds) Advancing banana and plantain R and D in Asia and the Pacific, vol 9. The international network for the improvement of banana and plantain-Aisa and the Pacific Network (INIBAP-ASPNET)
39. Warwicker JO, Hallam P (1970) 6—The effect of alkaline and acid swelling agents on the mechanical properties of cotton fibres. J Text Inst 61(2):61–76
40. Weerasooriya R (2016, July 15) Banana nation: your comprehensive guide to Sri Lankan Bananas. Sri Lanka Foundation International. Retrieved March 4, 2017, from https://www.srilankafoundation.org/newsfeed/banana-nation-your-comprehensive-guide-to-sri-lankan-bananas/

Achieving Sustainable Development Goals: The Impact of the Apparel and Textile Industry on the Eradication of Poverty in Sri Lanka

Thushari N. Vidanage, Lakshika C. Weragoda, Muditha Karunarathna, and Wasantha Athukorala

Abstract Poverty eradication is the first goal of the United Nations Sustainable Development Goals (SDGs) set to be achieved by 2030. As a developing country, Sri Lanka is committed to achieve these goals and to become a sustainably developed country. However, the COVID-19 pandemic and the deepening economic crisis have created a serious challenge in achieving these goals for Sri Lanka. In this context, the apparel and textile sector, one of the largest contributors to the economy, has a great role to play. The creation of direct and indirect employment and the increase in incomes and savings of households are among the significant impacts of the apparel and textile industry on the economy. However, a thorough assessment of the impact of the apparel and textile industry on poverty reduction proves challenging within the extant scholarly literature. This important gap sheds light on the aims of our study, which are to (i) investigate the impact of the apparel sector on the reduction of income poverty in Sri Lanka, (ii) explore the potential impact of the apparel and textile industry on eradicating multidimensional poverty in Sri Lanka, and (iii) examine the role of the apparel sector on fostering female empowerment in the country. This research is based on primary data, which is collected from employees in the apparel sector through a survey questionnaire. The sample size is 185 respondents who are workers in the apparel sector, covering large, medium, and small-scale apparel and textile firms island-wide. In addition, secondary data is collected from various sources. Based on the gathered data, a detailed quantitative and qualitative analysis is conducted to test the validity of the hypothesis that the apparel industry helps to reduce poverty in Sri Lanka. The findings indicate that over one-third of the

T. N. Vidanage (✉) · L. C. Weragoda · M. Karunarathna · W. Athukorala
Department of Economics and Statistics, University of Peradeniya, Peradeniya, Sri Lanka
e-mail: thushari.vidanage@arts.pdn.ac.lk

L. C. Weragoda
e-mail: lakshika.wrg@gmail.com

M. Karunarathna
e-mail: mudithakr@yahoo.com

W. Athukorala
e-mail: wathukorala@yahoo.com

G. M. Ranathunga (ed.), *Textiles of Sri Lanka*, SDGs and Textiles,
https://doi.org/10.1007/978-981-97-0569-6_5

sample lives below the income poverty threshold, with nearly 20% experiencing multidimensional poverty. Evidently, the apparel and textile industry significantly contributes to reducing both income and multidimensional poverty in the country.

Keywords Sustainable development goals · Apparel and textile industry · Sri Lanka · Poverty · Female empowerment

1 Background to the Study

In 2015, the United Nations, representing all its member states, unanimously proposed the 2030 Agenda for Sustainable Development. Within this agenda, there are 17 Sustainable Development Goals (SDGs) aimed at being realized by 2030, with the first goal being the elimination of poverty in all its forms everywhere. Poverty in Sri Lanka has doubled in the past two years from about 13–25%, causing an additional 2.5 million people to fall into poverty. Urban and rural poverty has tripled and doubled in 2022, to 15 and 26%, respectively [42]. This rise in poverty during 2021–2022 has also led to further declining multidimensional deprivations such as reduced investments in health, lack of nutrition, and disruptions to education, especially in the estate sector where the majority of the estate community live below the $ 3.65 poverty line (see Fig. 1). However, over the long term from 2002 to 2019, a decline is observed in Sri Lanka's poverty from about 47% to 14%, respectively, using the updated poverty line [10, 11].

With this background, Sri Lanka, as a developing country, shares in the commitment to achieve the SDGs set by the United Nations and upgrade itself as a sustainably developed nation. In its path toward achieving SDG 1, the apparel and textile sector is of paramount importance. It plays a vital role in the Sri Lankan economy by contributing to generating both direct and indirect employment, export revenues, and foreign exchange earnings. The sector also serves as a key player in economic

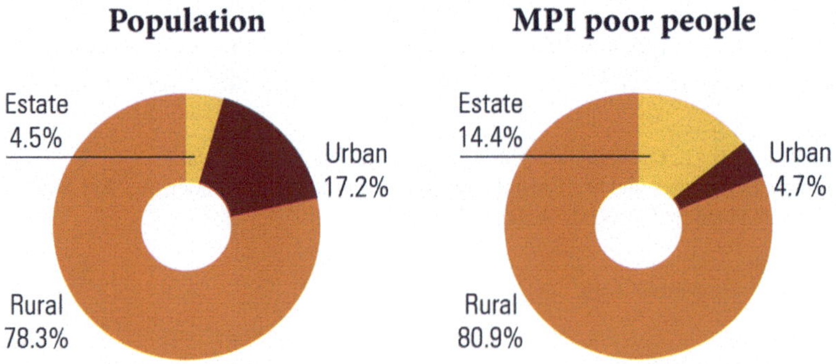

Fig. 1 Distribution of population and the multidimensionally poor in different sectors (2019). *Source* Department of Census and Statistics

development, especially in rural areas, and has contributed to Sri Lanka's efforts to reduce poverty and empower women in the workforce.

During the last 10 years, the apparel and textile sector has contributed to around 45% of Sri Lanka's total merchandise exports and around 50% of the industrial product exports [9]. Of the total exports manufactured by firms in the Board of Investment (BOI) in 2021, the apparel and textile sector products accounted for 60% [7]. Although the COVID-19 pandemic and the country's economic crisis have adversely led to reduced export earnings toward the latter part of 2022, Sri Lanka's apparel and textile sector continues to be the dominant export earner. Among the exporting countries in the region, Sri Lanka stands with the highest exports per capita [36].

While there is limited literature on this specific topic in Sri Lanka and overseas, several studies have touched upon related aspects such as the industry's contribution to employment, income generation, and socioeconomic development [5, 25–28]. Following previous research, the primary parameter of eradicating poverty in this study is direct employment generation [3, 21, 23, 24, 29, 32, 33, 39].

Apart from direct employment generation, following [1, 23], we also focused on the following: wages and savings of employees who are mostly low-skilled and drawn from the rural sector, education and healthcare of workers; skills and training that enhance their working potential in other industries; work culture that empowers workers, especially women who are predominantly employed in this sector; benefits to the households; and physical infrastructure. While the existing literature recognizes the important role of the apparel and textile sector in Sri Lanka and other countries in contributing to employment, income generation, and gender empowerment, further research is needed to provide a more comprehensive understanding of the industry's direct and indirect impacts on poverty reduction, particularly in the Sri Lankan context. Therefore, this study attempts to fill this research gap.

Accordingly, the current study focuses on the role of the apparel and textile sector in poverty eradication in Sri Lanka, undertaking a comprehensive examination of the sector's influence on addressing both income and multidimensional poverty, along with its role in fostering Sri Lanka's socioeconomic development. The primary objectives of the chapter are (1) to examine the impact of the apparel and textile sector on reducing income poverty in Sri Lanka; (2) to explore the potential impact of the apparel and textile industry on eradicating multidimensional poverty in Sri Lanka; and (3) to examine how the apparel and textile industry might foster female empowerment within Sri Lanka. Both primary and secondary data are used for the analysis. The results will provide policymakers with insights into the influence of the apparel and textile industry on poverty eradication in Sri Lanka.

The subsequent sections of this chapter are organized as follows. In the next section, the historical overview of the apparel sector is given. In Sect. 3, we delve into the importance of the apparel and textile sector in relation to poverty, drawing upon insights from previous studies. Section 4 analyzes the role of the sector in fostering female empowerment, backed again by previous research. Section 5 describes the data sources and methodology used in the study, while Sect. 6 analyzes the empirical findings. Finally, in the concluding section, we offer our concluding remarks.

2 Historical Overview of the Apparel and Textile Sector in Sri Lanka

In 1977, trade liberalization in Sri Lanka brought about a series of economic reforms that opened up the Sri Lankan economy to the global market. This move had a profound impact on the apparel and textile sector, leading to its emergence as a key player in the country's economy. The liberalization policies in 1977 resulted in removing many trade barriers and import restrictions, creating a favorable climate for exports. As a result, the apparel and textile industry began to flourish, initially driven by local entrepreneurs who saw opportunities in the export market. In the consecutive years following these reforms, Export Processing Zones (EPZs) were established that offered incentives such as tax benefits and favorable regulations to attract foreign investment in the sector. This reform, in particular, led many international clothing brands to outsource production to Sri Lanka due to cheap labor costs in the latter. During this period, Sri Lanka's apparel sector experienced robust growth, becoming one of the country's largest export industries.

During the 1990s, the apparel and textile industry diversified its product range, moving beyond simple garments to more value-added products. Sri Lanka was trying to build a reputation for producing high-quality apparel, which in later years helped attract premium brands and secure its position in the global supply chain. In the meantime, female labor force participation in the sector grew significantly, which led to many women becoming economically and socially empowered.

Over time, the termination of the Multi-Fiber Arrangement (MFA) in 2005 posed new challenges for Sri Lanka as quotas on textile and apparel exports were lifted. Sri Lanka faced increased competition for its exports, particularly from other Asian countries due to declined prices. Compared to other industries, the wage premium in the apparel and textile sector reduced with the dissolution of the MFA and the male–female wage gap widened [35]. However, the industry adapted by focusing on quality, innovation, and value addition to remaining competitive in the global market. It continued to grow since the late 2000s and shifted its focus toward sustainability and ethical practices, which has attracted environmentally conscious consumers and brands in the global market. Due to its experience with COVID-19, the Sri Lankan apparel and textile sector is recognized today for its resilience, adaptability, and innovation. Figure 2 shows the contribution of apparel and textile exports to total merchandise exports in the country between 2013 and 2023.

Over the years, men's and women's undergarments and outerwear continue to be the primary apparel and textile categories exported from Sri Lanka, while activewear/sportswear, babies' garments, and warm clothes account for a lesser share of the total exports (see Fig. 3). The highest export earnings of about US $ 3100 million in 2022 were brought by knitted or crocheted apparel and textile accessories, while non-knitted or crocheted articles brought in around US $ 1800 million [13]. The USA has historically remained the largest market for Sri Lankan apparel and textile exports accounting for earnings of nearly US $ 2500 million in 2022. Along with the USA, other European countries such as the UK, Italy, Germany, and the Netherlands

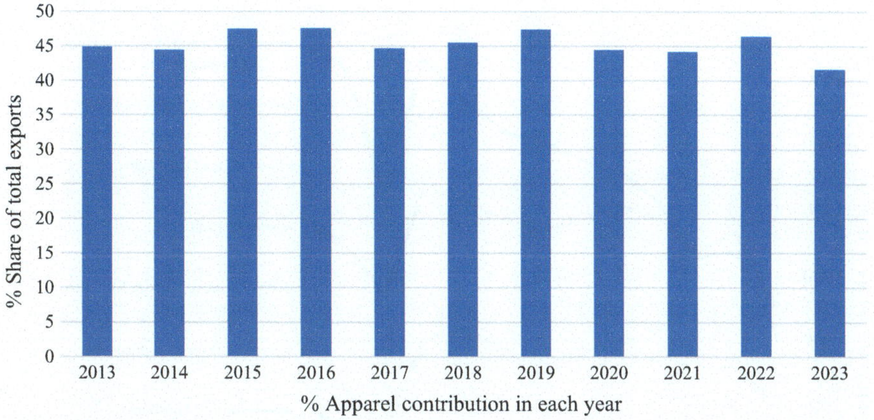

Fig. 2 Apparel and textile contribution to total merchandise exports (2013–2023). *Source* Export Development Board

together account for more than half of the total export earnings, particularly in 2022 (see Fig. 4).

The apparel and textile sector in Sri Lanka is entirely privately owned, and there are about 350 manufacturing firms in the country of which 10 are situated in the North and East. Around 350,000 people are employed in the industry as direct employees, while another 600,000 are provided indirect employment [13]. The majority, i.e., about 78% of the industry's workforce at present is female [4]. Among the projects approved by the BOI in 2022, the highest estimated employment capacity of around 6500 was recorded by the textile and apparel sector [37]. As such, it becomes imperative to ascertain the industry's role in alleviating poverty within the Sri Lankan context.

The role of the apparel and textile industry in Sri Lanka and other countries in eradicating poverty has been an area of academic and policy interest. The research findings of [18, 43] demonstrate that the apparel and textile industry has successfully alleviated poverty in Bangladesh and Cambodia, respectively, due to the provision of income-generating opportunities, especially for unskilled and semi-skilled workers and women. Where 50% of rural households live in poverty in Bangladesh, workers in the export-oriented apparel and textile sector were able to raise their living standards above the poverty line by being able to fulfill their basic needs with their earnings. In Cambodia on the other hand, entry-level workers in the apparel and textile sector earn much more than the poverty line.

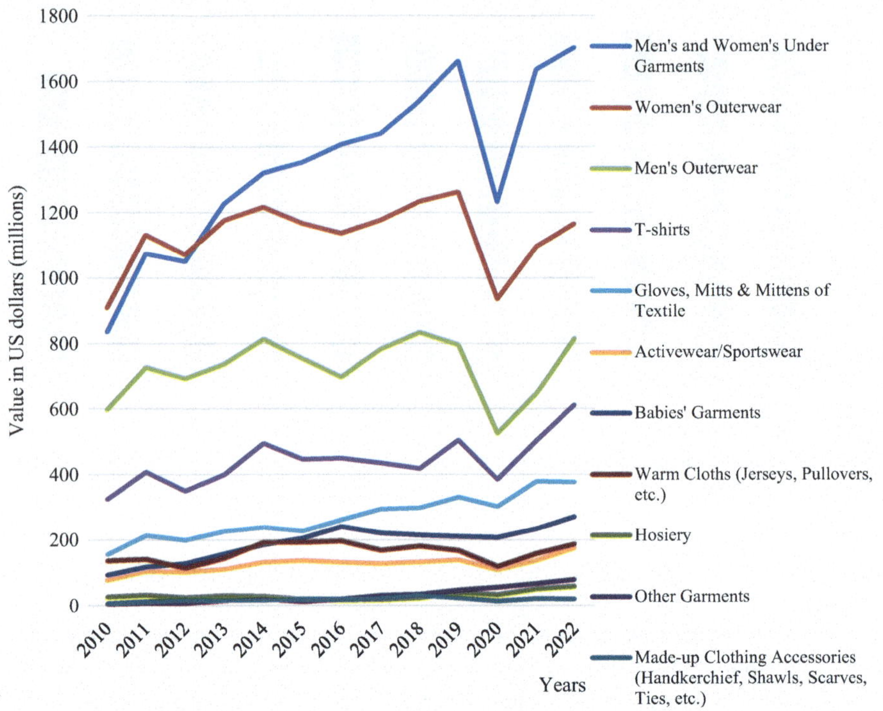

Fig. 3 Categories of apparel and textile products exported by Sri Lanka 2010–2022. *Source* Authors' compilation using data from the Export Development Board

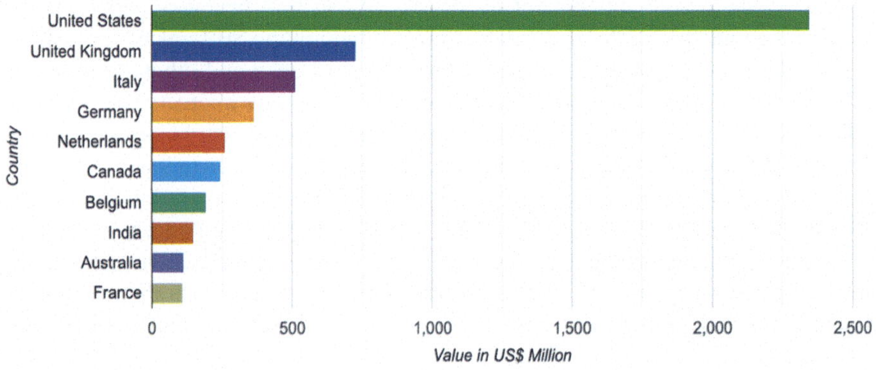

Fig. 4 Major export destinations of the apparel and textile sector, Sri Lanka—2022. *Source* Export Development Board

3 Importance of the Apparel and Textile Sector on the Reduction of Poverty

First and foremost, we aim to explore the impact of the apparel and textile sector on the reduction of income poverty in Sri Lanka. The sector's influence on income poverty is significant, particularly in Sri Lanka where the industry is a key driver for economic development. The apparel and textile industry in Sri Lanka is labor-intensive, thus employing a large number of people, especially in rural areas where there are limited economic alternatives. These employment opportunities range from factory workers to managerial positions and contribute to an increase in the household income of the workers, thereby reducing income poverty. The apparel and textile industry often employs low-skilled and semi-skilled workers, which makes the industry known to be hiring cheap labor. While wages in this industry are low, they can still provide a stable source of income for workers and their families, helping to lift them out of extreme poverty.

This is especially so when the apparel and textile sector can provide additional sources of income for families engaged in other income-generating activities such as agriculture, contract labor, or other low-income temporary employment. In many cases, we observed that one family member works in the apparel and textile sector, while others engage in different income-generating activities, helping to diversify household income. Overall, as Sri Lanka primarily exports apparel and textile products, the sector can bring in significant foreign exchange earnings. This income can again be used to invest in infrastructure, education, and healthcare, thus indirectly benefiting the population as a whole and reducing income poverty. Moreover, the growth of the apparel and textile industry can contribute to economic growth and create a spillover effect throughout the economy. As the industry expands, it generates increased economic activity, which increases income levels and reduces poverty.

Second, we seek to delve deeper into the potential impact of the apparel and textile industry on the eradication of multidimensional poverty in Sri Lanka, recognizing the complex and interrelated nature of poverty. The impact of the apparel and textile sector on multidimensional poverty is significant as it extends beyond income poverty and includes various aspects of standard of living and well-being. Being employed in the apparel and textile sector not only contributes to higher incomes but also offers access to social protection such as the Employment Provident Fund, Employment Trust Fund, and health benefits for workers, which in turn address other dimensions related to basic needs.

On the other hand, some apparel and textile firms provide leadership training and skill development programs for their employees. This is an investment in human capital that can address other dimensions related to education and empowerment. Most apparel and textile firms provide other social services for employees such as food and nutrition programs, medical leave, and childcare facilities, which again address health and nutrition-related dimensions. Employment in the apparel and textile sector and the industry's expansion in the country benefit the housing and living conditions of workers through the purchase of new household equipment and infrastructure improvements within the households. In addition, apparel and textile firms that value labor rights and ethical practices contribute to favorable working conditions, reduced discrimination, and a safer, more dignified work environment, which address aspects related to human rights and dignity.

A study by Islam et al. [21] emphasizes the role of the apparel and textile industry in creating employment opportunities in Bangladesh. It found that the sector provides jobs to a significant portion of the country's workforce, which includes a substantial number of illiterate female and rural workers. Maliwichi et al. [29] and Uddin and Jahed [39] emphasize the industry's potential for generating income for the poor. They note that the wage levels in the sector, though not uniformly high, are often above the low-wage economic activities in rural areas and contribute to improving people's living standards.

Labor-intensive industrialization as a pro-poor development strategy is highlighted by UNIDO [40]. Karunaratne and Abayasekara [23], Keane and Velde [24], and Weerasinghe [41] delve into the role of the apparel and textile industry in poverty reduction and growth and development strategies, particularly related to rural development. They argue that the establishment of apparel and textile factories in rural areas has led to improved infrastructure and increased income levels, which contribute to poverty reduction in these regions. In addition, strategies such as promoting export diversification, attracting foreign direct investments, and other applicable policies and interventions are implemented at the national level. On the other hand, [32] highlight that the industry's success in international markets has contributed to foreign exchange earnings and economic growth, with spillover effects on poverty reduction.

4 Impact of the Apparel and Textile Sector on Fostering Female Empowerment

Next, we aim to investigate how the apparel and textile industry could influence fostering female empowerment within Sri Lanka. The sector plays a vital role in fostering female empowerment in various means [19, 27, 38]. In the Sri Lankan context, the industry employs a largely female workforce, enabling gender inclusivity in the workplace. Women who work in the sector earn an income, which is a source of empowerment through financial independence and helps reduce gender-related poverty. Financial independence also enables women to control their economic resources, thus reducing their dependency on male members of the household and strengthening their decision-making power within households, especially related to children's education and health.

The leadership and skill development training that female employees receive at their workplaces equips them with skills that can also be applied in other areas of their lives. Many female employees use the income earned from their employment for the education of their children or younger siblings, which is a crucial factor for empowerment. Educated children in the next generation are more likely to break the vicious cycle of poverty.

It is important to note that issues related to gender discrimination and poor working conditions continue to be seen in some segments of the apparel and textile industry. Therefore, for female empowerment to be fully realized, the sector needs to focus on ensuring safe and fair working conditions, eliminating gender-based discrimination, and promoting a favorable work environment that supports women. The signature phrase of the Sri Lankan apparel and textile sector '*Garments without Guilt*' is a proactive positive measure in this regard, which improves the industry ethics and working conditions throughout the country [36].

Hancock et al. [17] examine the influence of the apparel and textile industry on gender empowerment. The study suggests that the industry has played a fundamental role in expanding employment opportunities for women and enhancing their economic participation, thus leading to improvements in gender equality and poverty reduction. In contrast, the authors also highlight that there are several disempowering factors such as harassment, long working hours, and poor nutrition that women need to deal with at the factory and community levels. In another study, [16] find that the capital accumulated by women in the workplace is further invested in economic and social capital at the individual and family levels. Renovations in the household, children's and young siblings' education, jewelry and household equipment, and medical expenses of the family members are some investments carried out by female employees [16, 23].

While acknowledging the positive impacts of industrial work on women's empowerment and the favorable shift in traditional gender roles, both of which primarily result from paid employment as the first step out of poverty, [6, 12, 15, 34] highlight that there are negative repercussions on women's health as a result of working for long hours in unsafe working conditions in industries. Another strand of literature identifies that occupational segregation and wage gaps continue to dominate the apparel and textile sector although there are vast improvements in narrowing down gaps in male–female employment, social recognition, bargaining power in the household, and control over economic resources [31].

5 Description of Data and Methodology

To achieve the objectives of this study, we largely use primary data, which is collected directly from employees within the apparel and textile sector through a survey questionnaire. Our sample size is 185 respondents. Questionnaires were sent to employees representing large, medium, and small-scale[1] apparel and textile firms in Sri Lanka, which were selected through purposive sampling regardless of whether they were

[1] We followed the same categorization of the EDB, where apparel and textile firms are categorized based on their annual export turnover as follows: above Rs. 750 million (large); between Rs. 16 million and Rs. 750 million (medium), and below Rs. 16 million (small).

export-oriented or domestic market-oriented. The questionnaires obtained quantitative data on employment, wages, and basic demographic and socioeconomic information. We surveyed firms that were inside the EPZs as well as outside the EPZs. Secondary data was also collected from several sources, including governmental reports, academic literature, and institutional data from the available records of the Export Development Board (EDB), Department of Census and Statistics (DCS), Central Bank of Sri Lanka (CBSL), Board of Investment (BOI), and websites, among others.

Our study exploits a quantitative and qualitative analysis to rigorously assess whether the Sri Lankan apparel and textile sector is a significant economic force in helping to reduce poverty within the country. This methodology is in line with previous research that focused on the impact of the apparel and textile industry, particularly the EPZs, in alleviating poverty [23, 43]. In terms of wages in particular, we analyzed how the wages of apparel and textile sector workers, over the years, have contributed to pulling the workers above the national poverty line. Previously, similar analyses concerning the poverty line have been carried out in Bangladesh and Cambodia [18, 43]. While the analysis of wages and the poverty line helps us achieve the first and second objectives of our research, a sub-analysis contributes to achieving the third objective.

6 Results and Discussion

6.1 Basic Characteristics of the Sample

This section intends to provide basic characteristics of the sample, including the age, gender, education level, family background of workers, and firm size. It will be followed by a detailed analysis of income poverty in Sect. 6.2, multidimensional poverty in Sect. 6.3, and female empowerment in Sect. 6.4.

The age structure and gender distribution of the sample are given in Fig. 5. The age structure of the sample (Fig. 5a) reveals that the majority of the workers (51%) are aged less than 30 years, while 41% are between 30 and 50 years of age. The female workers dominated the sample as 83% of the respondents were female (Fig. 5b). The sample suggests that relatively young females work in the apparel sector in Sri Lanka.

The apparel and textile sector emerged in the country as a solution for the rising youth unemployment, particularly among females in the 1980s. Machine operators and related positions in this sector do not require high educational qualifications such as university degrees or skilled vocational qualifications. Thus, those who have left school education with the G.C.E. Ordinary Level (O/L) Examination or earlier are considered ideal to take up these jobs in the apparel and textile sector. This situation is evident in the surveyed sample. As shown in Fig. 6, more than 78% of the workers in the sample possess secondary school qualifications, while 19% of the

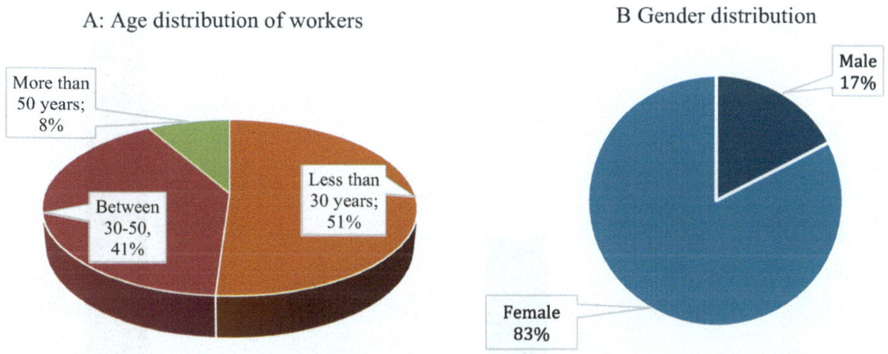

Fig. 5 Age and gender of workers (**a**, **b**). *Source* Authors' calculations based on survey data

workers have only attained primary school education. Notably, none of the workers in the sample have university or higher degree qualifications. Moreover, many of these workers work in the same district they resided in before commencing their employment, while a few have relocated to another district to undertake their jobs. This suggests a certain degree of labor mobility within the country.

Concerning the family background of the workers in the sample, it is evident that the majority of their families (126) comprise at least 3 or more family members, and many of them live with extended families. This indicates that the majority of members are still breadwinners in their families, suggesting that only one or two members in their families generate some form of monthly income. It is noteworthy to observe that certain family members have either completed or are currently pursuing higher education such as a university degree. As previously mentioned, this is a significant observation, considering that many of these surveyed workers have not passed at least the G.C.E. Advanced Level (A/L) examination, suggesting a relatively limited educational background to undertake *white-collar* jobs. Nevertheless, some family

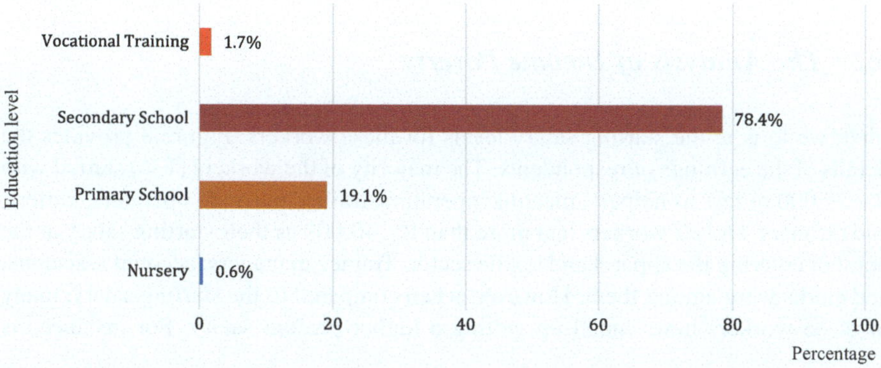

Fig. 6 Education level of workers. *Source* Authors' calculations based on survey data

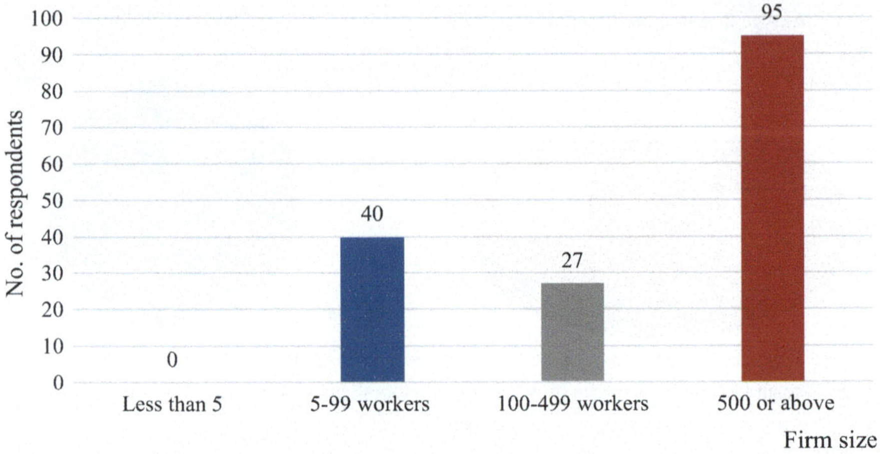

Fig. 7 Size of the firm in the surveyed sample. *Source* Authors' calculations based on survey data

members of these workers, i.e., their children or siblings, have been able to attain higher educational qualifications as a result of the workers securing employment in the apparel and textile sector. Possessing these higher educational qualifications would give the former potentially better high-income earning opportunities in the future, thus gradually diminishing the burden of family commitments on apparel and textile workers. In addition to the above family commitments, at least a quarter of the workers in the apparel and textile sector are also faced with relatively high medical costs in the family budget as a result of chronic illnesses prevalent among their family members, the latter depicting their inability to work for prolonged periods.

As illustrated in Fig. 7, the majority of the workers work in large-scale factories[2] where the number of workers exceeds 500. Nearly 25% of workers belong to middle-scale and small-scale factories.

6.2 *The Analysis of Income Poverty*

First, we look at the starting salary levels for these workers. Figure 8 provides the details of the earnings of respondents. The majority of the workers (136) started with Rs. 25,000 or less as helpers, machine operators, trainee machine operators, cutters, and cleaners. Only 7 workers had more than Rs. 40,000 as their starting salary at the point of entering the apparel and textile sector. Trainee managers, account assistants, and clerks were among them. However, when compared to the starting salary, many of these workers have ended up getting a higher present salary. For instance, as

[2] See the scales for the size of the firms at https://www.investopedia.com/terms/s/smallandmidsize enterprises.asp retrieved on 10-11-2023 [20].

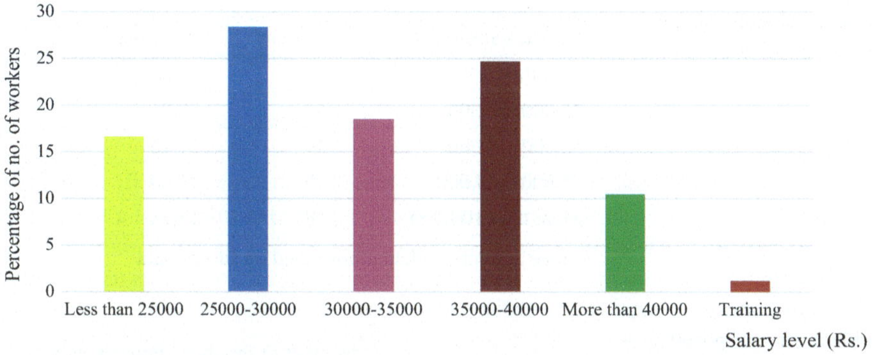

Fig. 8 Present salary levels of workers (in Rs.). *Source* Authors' calculations based on survey data

illustrated in Fig. 8, only 27 (16.6%) workers have a salary of less than Rs. 25,000. Another 28% have a monthly salary between Rs. 25,000 and 30,000. The majority of them have been promoted to the positions of machine operator, team member, or team leader. About 43% of the sample receive a monthly salary of Rs. 30,000–40,000. Nearly 10% of the sample earns more than Rs. 40,000, and this category covers more of the white-collar jobs in the sector.

The first main objective of the present chapter is to examine the role of the apparel and textile sector in reducing income poverty in Sri Lanka. The information on the income of the workers and their family members for a particular month was collected and confirmed through a variety of questions in the survey questionnaire. Based on the National Poverty Line (NPL) constructed by the Department of Census and Statistics for December 2022, the monthly income poverty thresholds for each family were determined.[3] The NPL for December 2022 was calculated as Rs. 13,777 per head. Table 1 shows the family income distribution of the workers in the sample based on the information given for September 2023. The monthly family income is less than Rs. 50,000 in nearly 33% of the families (one-third of the total sample). For most of the families in this category, the main income source was the salary earned by the apparel sector worker. Without the income derived from the apparel and textile sector, these families would have faced significant challenges in meeting their daily needs and ensuring survival. In contrast, another 48% of the sample (27 + 21) lies between a family income ranging from Rs. 50,000 to 100,000 where several family members under these two categories significantly contributed to the total monthly family income so that the burden on the apparel and textile worker was relaxed. Table 1 provides further details about the income distribution of the sample.

As the next step based on the above calculations, poverty ratios were calculated. As illustrated in Table 2, 34.5% of the surveyed sample could be categorized as

[3] The NPL for December 2022 is the latest, which we used to calculate the income poverty thresholds for the month of September, 2023. However, due to lack of data, we were unable to construct the poverty lines for the previous periods. Thus, a possible comparison has to be ruled out.

Table 5 Perception of women's empowerment among female workers

Category	Total number of responses	Strongly agree (%)	Agree (%)	Neutral (%)	Disagree (%)	Strongly disagree (%)
(i) I have control over my income and expenses	108	11.7	55.4	30.2	1.6	0.8
(ii) I have seen changes in my social status as a result of working in the garment industry	109	4.5	55.9	37.6	1.8	0
(iii) I feel more capable and confident due to my job	118	9.3	73.7	16.9	0	0
(iv) I feel safe and respected at my job	106	14.1	67.9	17.9	0	0
(v) I have experienced discrimination/ harassment in my workplace	96	0	7.2	15.6	58.3	18.7
(vi) I have experienced violence/harassment in my intimate relationships	101	0	9.9	7.9	59.4	22.7
(vii) I have the freedom to discuss and decide the use of contraceptives/ family planning methods	81	11.1	59.2	24.6	3.7	1.2
(viii) I'm involved in household decision-making on children's education/ health	79	15.1	63.2	18.9	2.5	0

Source Authors' calculations based on survey data

Another often-cited indicator of women's empowerment is their decision-making ability. We asked what the female workers perceived about their *involvement in household decision-making on children's education and health*. More than 60% of mothers agree with this statement, while another 15% strongly agree with it. Hence, we could conclude that at least the female workers who agreed with the statement in general believe that they are largely involved in the household decision-making process, which again indicates a positive sign of female empowerment. The increased financial independence as a result of their job in the apparel sector is highly likely to have significantly improved their voice in the household. In other words, employment in the apparel sector has transformed women from voiceless individuals to forceful decision-makers at the household level.

The majority of female workers indicated that they have never been engaged in leadership roles or decision-making processes at their workplace. Therefore, we

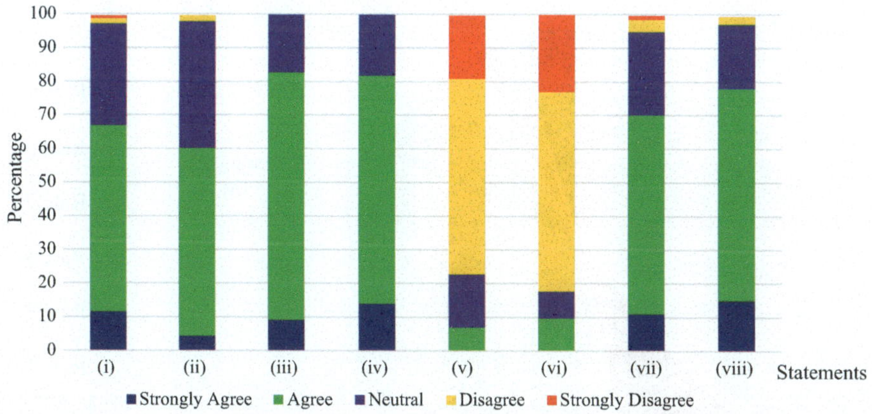

Fig. 10 Perceptions of female empowerment. *Source* Authors' calculations based on survey data

could suggest that among the women surveyed, their involvement in household decision-making surpasses their participation in workplace activities. This could be because most of the female workers in the apparel industry are in the positions of machine operators or work in similar capacities, thus engaging in minimum management roles. A similar conclusion was obtained regarding their involvement in community or political activities related to work.

As the last step of this analysis, female respondents were requested to provide a rating for the overall sense of empowerment. More than 80% of the respondents rated *8 or higher* as their perception of overall empowerment (see Fig. 11). Among them, nearly 40% are *highly satisfied*, i.e., marked a rate of 10. The minimum rating of 5 was reported by three respondents in the female sample. None of them had marked *dissatisfied*, i.e., a rating of 1–4 on the scale.

The preceding analysis on female empowerment implies that the apparel and textile industry has a crucial role to play in fostering female empowerment within Sri Lanka. Previously, we documented that this sector has created a significant impact on the reduction of income poverty in Sri Lanka. Presently, it appears that employment opportunities in the apparel sector have empowered these workers, especially women, to attain a level of self-confidence, worthiness, and professional maturity. Many have enhanced their access to opportunities and resources, benefiting not only themselves but also their siblings and children. Their decision-making and bargaining power, particularly at the household level, has improved, which provides them the capacity to effect positive changes in their lives.

It has to be noted that some female respondents refrain from answering the above questions on empowerment (nearly 10% of the total female respondents). It is difficult to understand their sentiments and why they have done so in this particular section. On the contrary, despite the positive indicators of empowerment among female workers in the apparel sector that we have observed, there remains ample room for improvement. A notable weakness is the limited experience in the workplace decision-making

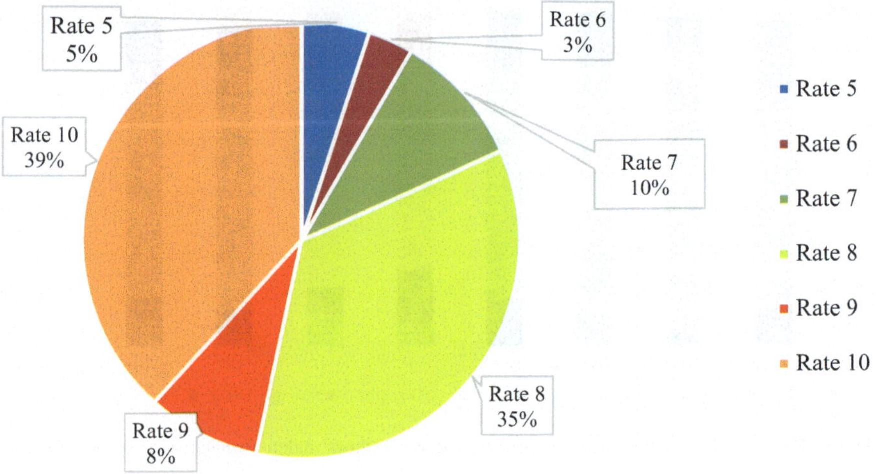

Fig. 11 Overall sense of empowerment as a female worker in the apparel industry. *Source* Authors' calculations based on survey data

process, particularly for those who commence their careers as machine operators who often end their careers in similar positions. Given the highly technical nature of present-day apparel industry operations, many decisions are made by higher-level managers, resulting in minimal involvement from the low-level workers in the decision-making process. Additionally, workplace discrimination and harassment are noteworthy issues. Extensive literature has thoroughly explored this aspect. Some female workers may choose not to disclose these personal challenges, potentially leading to underreporting of the actual situation.

7 Concluding Remarks

The objectives of the present chapter are threefold. First, we examined the impact of the apparel and textile sector on the reduction of income poverty in Sri Lanka. Using the selected sample of households, we found that more than 33% (one-third of the sample) live under miserable poor conditions, as their monthly income level is less than the NPL poverty threshold. This implies that without the income earned from the apparel sector, these households would have to struggle hard to survive. As the rest of the families in the sample are above the income threshold, it implies how important is the earnings generated through the apparel and textile industry for their households. These figures saliently indicate that there is a significant impact from the apparel sector on the reduction of income poverty within the country. Next, we explored how the impact of the apparel and textile industry extended to reduce the multidimensional poverty in Sri Lanka. The calculated headcount ratio indicates that nearly 20% of the respondents are multidimensionally poor. Third, we investigated how the apparel and textile sector could potentially foster female empowerment in Sri Lanka. We found some strong positive indicators of female empowerment among these workers. Their decision-making ability is significantly improved at the household level, leading them to trust themselves and be confident in what they do. Yet, there are avenues to improve such as their decision-making power at the workplace. Besides, their leadership and organizing capabilities could be improved considerably. In conclusion, we emphasize that the hypothesis asserting the role of the apparel and textile industry in poverty reduction in Sri Lanka remains valid and cannot be refuted.

Among potential policy recommendations, emphasis could be placed on strategies for income redistribution and the implementation of policies targeting the reduction of multidimensional poverty. In addition, there is a vital need to empower female workers in the industry, recognizing that a highly empowered female workforce in the sector holds the key to unlocking new opportunities for growth in terms of efficiency, specialization, and innovation.

Acknowledgements The authors would like to acknowledge Godage Piyathissa, Krishan Suranga, Shyamali Liyanage, and Lanka Tharuk for their invaluable assistance with the distribution of the survey questionnaire to employees in apparel and textile firms. The authors' special thanks go to Hilmiya Angel for the Tamil translation of the survey questionnaire. Sheshan Madhusanka and Hilmiya Angel were excellent research assistants for part of the data tabulation of this study. The authors sincerely acknowledge the employees and managerial staff of the firms who kindly cooperated by responding to the questionnaires.

References

1. Abeywardene J, de Alwis R, Jayasena A, Jayaweera S, Sanmugam T (1994) Export processing zones in Sri Lanka: economic impact and social issues, No. 69
2. Ada Derana (2023) At least 500,000 Sri Lankans lost their jobs in 2022—a top World Bank official. https://www.adaderana.lk/news.php?nid=88715. Retrieved on 10 Nov 2023
3. Amjad R (ed) (1981) The development labor-intensive industry in ASEAN countries: an overview. Asian Employment Programme, International Labour Organisation, Bangkok
4. Asian Development Bank (2020) Securing women's place in Sri Lanka's apparel industry. https://development.asia/insight/securing-womens-place-sri-lankas-apparel-industry#:~:text=In%20Sri%20Lanka%2C%20the%20apparel,reduced%20demand%20for%20women%20workers.
5. Athukorala P-C (2017) Manufacturing exports from Sri Lanka: opportunities, achievements and policy options. Institute of Policy Studies of Sri Lanka, Colombo
6. Attanapola CT (2004) Changing gender roles and health impacts among female workers in export-processing industries in Sri Lanka. Soc Sci Med 58:2301–2312. https://doi.org/10.1016/j.socscimed.2003.08.022
7. Board of Investment of Sri Lanka (BOI) (2021) Annual report 2021. BOI, Colombo
8. Caldwell JC (1980) Mass education as a determinant of the timing of fertility decline. Population and development review, 225–255
9. Central Bank of Sri Lanka (CBSL) (2022) Annual report 2022. CBSL, Colombo

Keywords Post-consumer waste · Circular economy · Secondhand fashion consumption · Sri Lankan consumers · Quantitative research

1 Background

The fashion industry is a significant contributor to the global economy; however, at the same time, it consumes a large number of resources and causes adverse environmental and social impacts that inevitably occur during a garment's production, consumption, and end-of-life phase [1, 2]. Fashion and apparel production accounts for 8% of greenhouse gas emissions. Because manufacturing is concentrated in Asia, the industry generates energy and heat from hard coal and natural gas. According to Elisa Tonda, Head of the UN Environment Program's Consumption and Production Unit, if we continue our current course, greenhouse gas emissions from industry will increase by about 50% by 2030.

As a result, the fashion industry will negatively lead to global warming, compounding the challenges that the Paris Agreement will confront in achieving its primary goals, one of which is to control global warming under 1.5 °C in the remaining years. However, this goal will be missed if the fashion industry continues to pollute the environment at its current levels. The speed of advancement in achieving sustainability in the fashion sector has dropped by a third compared to the previous year, according to the 2019 Pulse of the Fashion Industry report. The research also claims that the industry is not moving fast enough to counteract the sector's significant impact of rapid growth.

Every human being is a participant in the consumption of products marketed by the fashion industry. Naturally, this has impacted the environment, causing pollution from the pre-consumer stage to the post-consumer stage. As previously studied, pre-purchase, purchase, and post-purchase components are all part of consumption resulting from consumer behaviour. Purchasing goods others have already used is secondhand consumption [3]. But presently, consumers are concerned about using goods already handled by others, mainly due to the health risks posed by the pandemic. During the COVID-19 global pandemic in Europe, overflowing bins collecting textile waste drew the public's attention. The pandemic greatly impacted reuse and recycling companies, severely limiting their ability to operate. Textile businesses could not gather used garments because citizens had more time to clean their closets. When looking at the same group of signatories, the total amount of clothing collected decreased from 121,000 tonnes in 2019 to 90,000 tonnes in 2020 [4].

2 The Research Problem

The level of environmental pollution in the fashion industry has not diminished but is increasing daily [5]. Hence, most countries are currently aware of post-consumer fashion waste; as a result, they are seeking solutions to reduce this devastating environmental impact. However, in the context of Sri Lanka, there is a substantial theoretical and empirical knowledge gap concerning secondhand fashion consumption. Hence, the extant research seeks to fill the identified theoretical and empirical gap. To examine the impact of Sri Lankan consumers' value perceptions on consumers' intention to purchase secondhand fashion products and their word-of-mouth intention towards secondhand fashion products. This research aligns with the 12th SDG sustainable development goals (SDGs)-Ensure sustainable consumption and production patterns.

2.1 Literature Review

3 Environmental Destruction Occurred in the Fashion Industry

The apparel industry's global sale and production of 80 billion to 150 billion clothes annually contributes to further environmental harm. The fashion industry is responsible for 10–8% of the world's greenhouse gas emissions, which is higher than the maritime shipping and aviation sectors combined. The most significant environmental concern of the time is due to the previously listed factors in the fashion sector (see Fig. 1).

Within years after production, over three-fifths of all apparel is disposed of in landfills or incinerators. There are growing negative things about sustainability issues within the fashion supply chain due to the detrimental environmental effects of current fashion consumption habits [6]. A garment may be discarded for several reasons.

Fig. 1 Environmental destructions occur by the fashion industry. Author (2023)

The customer might no longer be interested in keeping the item since it is out of fashion. Other factors contributing to garment disposal include poor fit, quality, wear, and boredom [7]. In addition to depleting water resources and polluting rivers and streams, the fashion industry accounts for 10% of all human carbon emissions.

Additionally, 85% of all textiles end up in landfills each year. Thousands of pieces of plastic are also released into the water when washing various types of clothing [8]. Furthermore, the fast-fashion business impacts customer purchasing habits and encourages frequent garment disposal. The fast-fashion spread due to more affordable, quicker means of manufacturing and shipping and a greater demand for contemporary fashions among customers [9]. Leading apparel retailers like Zara and H&M have embraced fast fashion by bringing new clothing designs to stores every three to five weeks. So, customers are encouraged to regularly replace the clothes in their wardrobe with trendy items [10].

Impulsive buying and rapid fluctuations in demand are just two of the many issues that have caused overproduction in the fashion business; the issue is not just the overproduction of clothing; it also includes what occurs after the product is "consumed," with one-third of all clothing ending up in waste [11]. Landfilling is the leading way to dispose of garbage in most nations, and the type of fibre used to make the textiles affects how they fare in terms of environmental impact. Natural fibre-based textiles made of cotton and bamboo will eventually decompose, but like other biodegradable materials, they may produce pollutants, including methane and hydrogen sulphide [12].

4 Post-consumer Fashion Waste

In the United States of America, the largest consumer of fashion products is an estimated 92 million tonnes of post-consumer textile waste consigned to landfill sites every year [13]. Post-consumer garbage will skyrocket to uncontrollable levels as the world's population expands. From 8.3 million tonnes in 1999 to 11.3 million tonnes in 2009, the amount of post-consumer textile waste in the United States increased by 40% in just ten years, and the Council for Textile Recycling estimates that this number will expand even further to 16.1 million tonnes by 2019. Only 2% more rubbish has been diverted from landfills over the same period [14].

Due to the enormous consumer demand, fast fashion is a strategy in which the fashion industry offers products at a low price [15]. It led the fashion industry to a destructive contemporary way of producing and consuming. Several environmental issues are to blame for the fashion industry's irresponsibility and drive for more significant profit margins, the sector has become a leader and influencer in sustainability [16]. Thus, post-consumer waste accumulation will become a significant problem by causing environmental destruction through massive landfilling activities [17]. Post-consumer trash recycling garners more attention due to public awareness, legal advocacy and implementation, economic benefits, and the availability of suitable technologies. Allwood [18] drew attention to the importance of appraising post-consumer waste.

Various scientific investigations have found that fast-fashion's detachability and disposable methods have serious environmental, health, sociological, and economic effects [19]. According to the state news agency Xinhua, China overtook the US as the world's biggest fashion market in 2019 due to the popularity of e-commerce among the middle class. As a result, some countries produce 26 million tonnes of post-consumer garbage each year, of which only 1% is reused or recycled [17]. Developed nations have access to more reports on post-consumer textile and apparel waste than developing nations. Unfortunately, there are not enough statistics in Sri Lanka to determine how much consumers generate textile and clothing waste [13].

5 The Circular Economy in the Fashion Industry

The term "circular economy" was initially used to define a regenerative supply and demand chain (CE) [2]. CE tends to make a meaningful difference in economics because it endorses the practice of reusing goods that still have some life left in them, repairing things that are broken but can be reconditioned, and recycling those things that cannot be salvaged [13]. Importantly, this circular model shows the movement of resources through the recycling and resource recovery system (see Fig. 2). The circular economy has bridged the gap between manufacturing and consumption by recycling and reusing. The manufacturing stage will involve both the design and production activities of the circular economy, while the consumption will be controlled by using fewer material inputs to minimise waste. All the collected waste will be sorted, and the recyclable material will be separated. The waste dumping in landfills will be reduced by maximising the collection of reusable materials during the sorting stage which will have beneficial results by reducing the usage of natural resources. When this approach is adopted, valuable materials can be reclaimed by recycling, leaving only a limited amount of waste to be dumped in landfills. Thus, it will be possible to achieve the previously mentioned environmental benefits. Hence, clothing, footwear, and accessories must be planned, sourced, made, and offered to be securely returned to the biosphere after humans no longer require them to be utilised in society for as long as possible in their most valued form

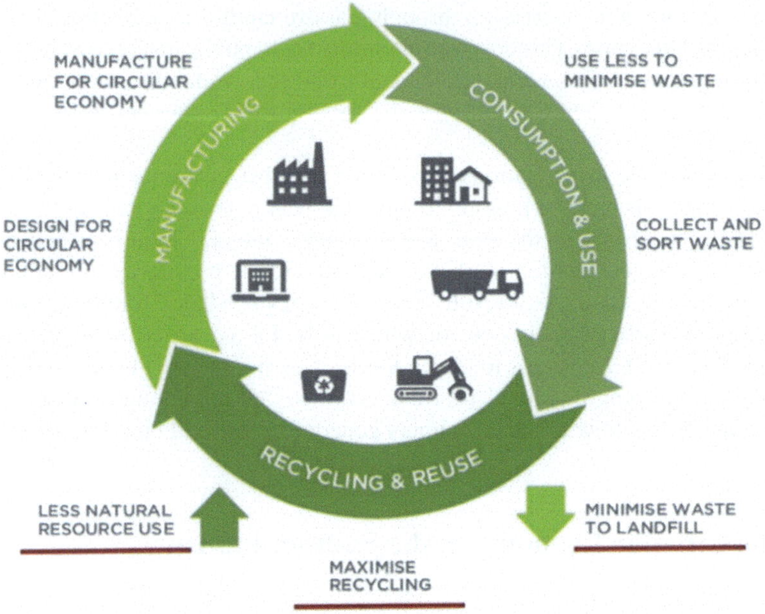

Fig. 2 Circular economy model for resources. *Source* Future Recycling (2018)

6 The Concept of Reuse and Fashion

The concept of reuse is already being practised in consumer industries. However, consumers need to be motivated sufficiently to broaden their environmental awareness to realise the importance of appropriately reusing products and disposing of waste materials [20]. In their research, [21] discovered three different findings on pro-environmental behaviour. Customers who recycle, for example, had previously spent time learning about environmental sustainability in their own time. Because they were already familiar with the strategy, customers who handed away products did not seek information from their colleagues to be reused, and they were inclined to acquire secondhand goods. Finally, clients who travelled by fewer polluting means of transport were more environmentally conscious. People started donating or selling unwanted garments to thrift shops, selling through alternative channels, or swapping among family and friends to dispose of them [22].

7 Secondhand Fashion Consumption

Fashion consumption is driven mainly by the need to keep up with current fashion trends at a breakneck speed, leading to short product life cycles and manufacturers' adoption of a fast-fashion marketing strategy [23]. Fashion clients' propensity to keep items inert and underused, simply filling closet capacity, sabotages efforts to pursue an environmentally friendly strategy. Researchers have noticed an increase in textile waste, dormancy in the usage of garments, and a faster speed of fashion consumption [24]. Over the last two decades, awareness of the negative environmental consequences has increased among fashion consumers. Fashion consumption also has been fuelled sustainably by efforts to reduce clothing waste. Although buying used fashion items is a green habit, consideration, and motivation caused by the quality or psychological attachments other than the environment have driven consumers to seek other sale platforms [25]. In recent years, secondhand fashion consumption has been quietly transforming. Thrift stores specialise in selling secondhand goods at rates often lower than those found within a free market [3]. Selling unwanted clothing to thrift stores, swapping it for other used clothing, giving it away to close relatives and friends, selling through unconventional venues such as flea markets, vintage stores, and online marketplaces, donating to charities, and recycling are just a few options for removing post-consumer waste

In the secondhand fashion business model, customers are the primary partners and suppliers to reduce resource consumption and waste [26]. According to [27], the SHC market grew twenty-one times greater than the retail sector in the last three years, and it is expected to increase to 51 billion dollars by 2023. The secondhand fashion trade has increased tenfold since 1990, but it still accounts for less than 0.5 cents of the overall terms of value, at around $1 billion per year. Secondhand fashion products sell for roughly 10–20% of the price of new clothes. Thus, the proportion is more significant in volume but still represents less than 5% of global trade. However, this proportion fluctuates dramatically depending on the receiving society and country. Almost every country is active in the trade, either as exporters, processors, re-exporters, or importers, with several countries performing multiple functions. Secondhand fashion products are primarily consumed in developing countries [28].

It should be mentioned that some observers believe the secondhand fashion industry in exporting countries is in trouble with the COVID-19 pandemic. Also, this is primarily due to an increase in the importation of low-cost, low-quality new apparel from Asia, which can be defined as the fast-fashion movement worldwide. However, import bans on used clothing are being implemented in some developing countries. Therefore, fast-fashion businesses may soon be obliged to collect more pre-owned things for reuse and recycling due to expanded producer responsibility rules [29]. According to fashion experts, customers must adapt their purchasing habits for the industry to survive. Be educated on the value of clothing, which requires collaboration between businesses and governments. So, there is an immense opportunity in retailing secondhand fashion products. Various industry experts say fashion resale will eclipse fashion retail in the next ten years [4].

Despite purchasing secondhand fashion products as an environmentally responsible habit, other psychological or material concerns may cause customers to look for other sales methods [25]. These elements may be linked to a customer's self-esteem and assets [30]. Therefore, consumer behaviour depends on consumer values.

8 Consumer Values

Consumers' purchasing behaviour plays a huge role in the fashion industry. Prior literature heavily shows that values are crucial for understanding consumer behaviour theoretically [31, 32]. This current research aims to understand consumers' perceptions of value towards secondhand consumption, is a barely-examined phenomenon theoretically and empirically. Additionally, Gutman (1990) discovered a link between individual values and intended customer advantages. A theory of values was created and empirically evaluated by Sheth in 1991. The theory was operationalised by defining five consumption values influencing consumer decision behaviour (Functional, social, emotional, epistemic, and conditional).

As a result, values govern all behaviour, including behaviours, attitudes towards people and things, ideologies, how one portrays oneself, evaluations, judgments, justifications, comparisons to others, and attempts to persuade others. Rokeach declared 1973 that "values are determinants of attitudes and action that are essentially or logically tied to them." Rokeach [33] noted that values "are guides and determinants of social attitudes and ideologies on the one hand and of social conduct on the other" when referring to the consequences of values. The notion that values significantly influence consumer behaviour is widely accepted [31, 32]. Values help describe consumer behaviour since they influence how customers behave regarding their choices of product categories, brands, retail locations, etc. [32, 34]. Customer values are essential because, as previously indicated, consumer behaviour determines consumer demand, which determines the fashion industry's production. Therefore, the values of the consumer are the subject of this study.

9 The Means-End-Chain Model and Consumer Behaviour

The Means-End Chain (MEC) model or the means-end theory has been used to study consumer behaviour [35]. The MEC model supports addressing and resolving the issues through quantitative methods by amalgamating the available qualitative data collecting and objective evaluation procedures [36]. Accordingly, this current research adopts the theoretical lens of the MEC model to explore the consumer's perception of value towards secondhand consumption. MEC deals with product attributes (A), consequences of product use (C), and personal consumer value (V) related to the use of the product [37].

The characteristics or features of the product are considered product attributes, which are divided into two categories. First, there are the concrete attributes or physical features such as the price, colour, size, materials. Then, there are abstract attributes and intangible product traits like quality, fit, and dependability [38, 39]. The advantages and disadvantages/risks arising from utilising a product or service in the normal course of life are the consequences (C). Three categories of effects have been discovered in previous investigations [38]. The first category includes the functional repercussions (FC), which is the experience of consumers who use the product. The second type covers the psychological consequences (PC), which refer to how the consumer feels about the features and attributes of the product [40]. The next type is the social consequence (SC), which is based on how the use of the product fits in with the social norms of others. A peak level of the desired end-state is defined as the value (V), a cognitive conception of the consumer's needs and goals [38].

10 Conceptualisation: Hypotheses Development

As informed by prior qualitative study findings of this research [13, 41], we have conceptualised ten hypotheses related to SHF consumption of Sri Lankan fashion consumers. Accordingly, both positive and negative effects were predicted (see Table 1).

11 Proposed Positive Effects on SHF Consumption

The previous study demonstrated the bond between secondhand fashion consumption and its financial benefits. When purchasing such things, consumers are driven by economic decisions. The consumer is encouraged to buy secondhand apparel since it is significantly less expensive than purchasing brand-new goods [48]. Therefore, consumers desire high-quality products at reasonable prices and consume secondhand fashion products for financial reasons [49]. Accordingly, the following ten hypotheses were formulated.

H1: Financial benefits will have a positive effect on consumers' intention to purchase Secondhand fashion products and their word-of-mouth intention towards Secondhand fashion products.

Findings further indicate the pleasure of the emotional state that consumers feel when they decide to purchase or consume secondhand fashion products. It demonstrates how admiration for the connection to or history of the previous owner and nostalgia are related [50]. Nostalgia is a craving for antique items that were once in style or a wish to dwell in a historical era that has since passed [51]. Customers are also thrilled with their dedication to environmental preservation and that they pick environmentally friendly clothing over harmful items. Making wise financial judgments is frequently described as being in a good mood [42]. As a result, emotional value is projected to impact the establishment of positive attitudes regarding circular fashion significantly. The following hypotheses were formulated.

H2: Emotional bonds with the secondhand products will have a positive effect on consumers' intention to purchase Secondhand fashion product and their word-of-mouth intention towards Secondhand fashion products.

Table 1 Operational table chart: hypothesis testing

Independent/dependent variable	Indicators/survey items	Source
Financial benefits	1. Buying used fashion items frequently offers excellent value for the money 2. Being able to purchase name brands that I otherwise could not afford new 3.Except for custom-made things, used clothing is typically less expensive than brand-new clothing 4. Cheap, but you can get some high-quality items in places that sell used clothing	[42]
Emotional bond	1. When I look at used clothing, I can tell right away that it is something I'd want 2. Secondhand clothing items would entice me to use them 3. Because the previous owner was someone important to me, I may utilise secondhand clothing items with confidence	[43]
Product quality	1. Usually, if I am specifically seeking for a used item, it's of high quality and will last for a long time 2. I am constantly searching for products that are virtually new, scarcely worn, clean, and made of high-quality materials	[43]
Self-satisfaction	1.Even though it's secondhand, wearing a particular brand may make you feel confident in yourself 2. I am quite proud of myself for buying used clothing 3. Less shame about not supporting quick fashion 4. If the item flatters you, it is good to receive praises from friends and let them know you're wearing a used item	[42]
Social status	1. I would feel ashamed to wear used clothing because I believe people would assume I could not afford to buy new 2. Some people I know may belittle others for these reasons or lack confidence in their financial management skills 3. My folks might worry about me and assume I'm having financial difficulties. Also, it appears that some people take great delight in their new purchases	[42]

(continued)

Table 1 (continued)

Independent/dependent variable	Indicators/survey items	Source
Hygienic factors	1. I adore shopping for used clothing. if they were consistently expertly cleaned and pressed 2. I detest the stench you get when you enter the store. The smell of old 'stuff' 3. There is no chance of contamination with secondhand clothing items	[42, 44]
Uncomfortability	1. I would prefer to know who owns the clothing 2. I just can't bring myself to put on anything that someone else wore and perspired in before me 3. There is a stigma associated with secondhand clothing that may suggest that the owner is uneducated or unable to purchase brand-new clothing	[42]
Self-expressiveness	1. Secondhand stores only provide items that individuals have donated, so your options are limited 2. People buy clothing to fit their own image; finding your sense of style in used clothing would be challenging 3. I have the impression of copying the prior owner. I do feel not special or original	[42]
Environmental consciousness	1. I care whether the products I use don't damage the environment 2. I worry about depleting the earth's resources 3. I consider myself to be someone who cares about the environment 4. The advantages of buying used clothing more frequently and the environmental advantages of doing so 5. I try my best to purchase clothing that is created from recyclable materials	[42, 45]
Purchase Intention	1. When I shop for apparel shortly, I'm willing to purchase used fashion items 2. If I am aware of the possible harm some clothing products can do to the environment, I avoid buying them	[46, 47]
Word-of-mouth intention	1. I want to share with others the experience and sensation of wearing used clothing 2. I would strongly advise it if someone asked my opinion on used fashion items	[46]

People who place great value on the environment lead and improve their lives as a result. They feel that beyond satisfying their physical or material wants, they should seek peace between nature and humankind [52]. As a result, they are very concerned and trying to be responsible for their purchasing and consumption behaviour. The circular economy encourages consumers to recycle and reuse materials to achieve a closed-loop supply chain [53, 53]. Therefore, environmental consciousness automatically encourages the consumption of secondhand fashion consumption.

H3: Environmental consciousness will have a positive effect on consumers' intention to purchase Secondhand fashion products and their word-of-mouth intention towards Secondhand fashion products.

Some customers also assume that vintage products are higher quality than newer items since they were built using more extraordinary materials. According to numerous research, secondhand fashion buying is a treasure-hunting hobby for discovering things that are of good quality and aren't widely available on the market [49]. Accordingly, the following hypothesis was formulated.

H4: Product quality will positively affect consumers' intention to purchase Secondhand fashion products and their word-of-mouth intention towards Secondhand fashion products.

Consumers behave differently based on their circumstances, which are influenced by financial circumstances, community involvement, and referrals from friends and family [42]. Helping charities by purchasing secondhand products, contributing to environmental conservation, and being humble enough to wear secondhand clothes lead consumers to self-satisfaction based on the development of this hypothesis [42].

H5: Self-satisfaction will positively affect consumers' intention to purchase Secondhand fashion products and their word-of-mouth intention towards Secondhand fashion products.

12 Proposed Negative Effect on the SHF Consumption

Social considerations usually have a significant impact on consumers' decision-making. They show their concerns about the image they project to their peers and worry about being judged when consuming secondhand fashion products. Consumers' most common concerns centre on the discomfort and stigma of being linked with persons of lower socioeconomic status [44]. Therefore, the following hypothesis was developed.

H6: Social status will negatively impact consumers' intention to purchase Secondhand fashion products and their word-of-mouth intention towards Secondhand fashion products.

Hygienic risk is the feeling of being anxious about something that could disturb one's health or life. Consumers explain that clothing, unlike other sorts of items, comes into direct contact with the body; hence, hygienic risk is likely to have an even more significant impact on consumer attitudes towards the product [46]. The idea is that circular fashion products are not clean and are likely to include stains or filth that could sicken the body when worn, as consumers in this study described. Even though the stigma associated with SHC is rapidly dissipating in Western countries [22], few prior studies found that some people still have a significant aversion to it due to its lack of cleanliness. As a result, cleanliness issues are a significant barrier to purchasing secondhand fashion products [54]. Accordingly, the following hypothesis was formulated.

H7: Hygienic consciousness of the products will negatively impact the intention to purchase Secondhand fashion products and their word-of-mouth intention towards Secondhand fashion products.

The relationship between people and their clothing can be complicated and intimate, with much subjectivity [55]. The fact that the items have been worn creates concerns among potential secondhand fashion buyers. The expected criteria are cleanliness and not knowing the previous apparel owners [22]. Therefore, the consumers explain that wearing secondhand clothing with the criterion is uncomfortable. Accordingly, the following hypothesis was formulated.

H8: Uncomfortability will negatively impact consumers' intention to purchase Secondhand fashion products and their word-of-mouth intention towards Secondhand fashion products.

To harmonise with customers' self-expression, demonstrating that apparel plays an essential role in aesthetically satisfying customers' desire for a similar self-image [56]. Similarly, worries about whether circular fashion clothing is appropriately coordinated with the consumer's other apparel, such as a lack of diversity, if the product complements the consumer's image, or whether the product fits poorly, are associated with the aesthetic risk of circular fashion clothing. For instance, buyers may be concerned that reusing fashion products lacks sufficient sizes, colour, and design diversity. They may also be concerned that the secondhand apparel worn and discarded by others is no longer fashionable [57]. Based on this reasoning, the following hypothesis was formulated.

H9a: Self-expressiveness will negatively impact consumer's intention to purchase Secondhand fashion products and their word-of-mouth intention towards Secondhand fashion products.

13 Research Design

This study took a quantitative approach [58]. Accordingly, a survey was conducted among a convenient sample of 257 participants in Sri Lanka with individuals aged between 20 and 50, the same as study one. The questionnaire contained 47 items. Section "A" of the questionnaire collected demographic information, consumers' purchase platform, consumer's awareness of SHF consumption, and their give-away and exchanging information. These questions were measured with yes/ no closed questions. Part B was formulated to test hypotheses. All the tests used five-point Likert scales with poles ranging from strongly agree to strongly disagree. A participant took approximately 10–20 min to complete the survey. The online survey was developed using Google Forms and administered via email and social media. Survey data was collected between 01/1/2021 and 01/01/2022.

14 Findings

14.1 Hypothesis Testing: Multiple Regression Analysis

14.1.1 DV: Purchase Intention

The results of the ANOVA analysis and if there is a statistically significant difference between variables are displayed in the table below (see Table 2). This table shows that the independent variables have a statistically significant difference.

It means that the significance value is 0.001, less than 0.05, and the mean duration of purchase intention among the independent variables on Sri Lankan fashion consumers' secondhand purchasing is statistically significant. The results can be stated as $F = 34.647$, $P \leq 0.05$. That is, the regression model is a good fit for the data.

The coefficient of variation allows you to compare variability among considerably distinct groups; that is, the findings of two systematic studies that do not use the same grading factors or value measurement (see Table 3).

According to the above table, the value of emotional bonds, self-satisfaction, hygiene factor, and uncomfortability was 0.245, 0.001,0.001, 0.003, and 0.013, respectively, considered statistically significant ($P < 0.05$). Additionally, the other independent variables (Financial benefits, environmental consciousness, product quality, special status, self-expressiveness) were insignificant. Therefore, the analysis confirms that emotional bonds of self-satisfaction as a model impact on increasing the efficiency in purchase intention. The beta coefficients of those variables were all positive, indicating that the independent and the dependent variables have a positive association.

Table 2 ANOVA table purchase intention

ANOVA[a]

Model		Sum of squares	df	Mean square	*F*	Sig.
1	Regression	222.354	9	24.706	34.647	< 0.001[b]
	Residual	176.129	247	0.713		
	Total	398.482	256			

[a] Dependent Variable: DV_Purchase_intention_1
[b] Predictors: (Constant)

Table 3 Purchase intention coefficient table

Model		Standardised coefficients	*t*	Sig.	95.0% confidence interval for B		Collinearity statistics	
		Beta			Lower bound	Upper bound	Tolerance	VIF
1	(Constant)		– 0.492	0.623	– 1.230	0.739		
	IV_financial_benefits	0.027	0.558	0.577	– 0.108	0.193	0.747	1.339
	IV_emotional_bonds	0.245	4.248	< 0.001	0.169	0.461	0.537	1.864
	IV_environmental_consciousness	0.078	1.565	0.119	– 0.041	0.358	0.716	1.397
	IV_quality_of_the_products	0.054	1.024	0.307	– 0.077	0.244	0.655	1.526
	IV_selfsatisfaction	0.370	5.280	< 0.001	0.328	0.718	0.364	2.744
	IV_social_status	– 0.039	– 0.778	0.437	– 0.203	0.088	0.719	1.392
	IV_hygienic_factors	0.147	2.988	0.003	0.090	0.440	0.743	1.347
	IV_uncomfortability	– 0.134	– 2.501	0.013	– 0.364	– 0.043	0.627	1.595
	IV_self-expressiveness	– 0.029	– 0.596	0.551	– 0.198	0.106	0.743	1.347

When the cleanliness component is increased by one unit and the other variables stay unchanged, the purchasing intention increases by 0.147 units. As a result, there is a link between raising purchase intention and increasing sales.

Regarding uncomfortability, the beta coefficient reveals a value of 0.013. It guarantees that if uncomfortability increases by one unit while all other factors remain constant, the purchasing intention will decrease by – 0.134 units. As a result, there is a negative correlation between discomfort and purchasing intention. The researchers expected that influencing factors like emotional ties and contentment substantially impacted purchasing intent. Uncomfortability also has a strong negative correlation with purchasing intention. The effect was shown as significant, and H2 was supported ($0.05 \leq p$-value), H5 was supported ($0.05 \leq p$-value), and H8 was supported ($0.05 \leq p$-value).

14.2 DV 2: Word-of-Mouth

The model summary results show that the correlation(R) is 0.610. The independent and dependent variables are linked this way (see Table 4).

The dependent variable is the correlation coefficient, and the independent variables correlate less than 0.7. The coefficient of determination (R^2) is 0.624, and the percentage of the dependent variable covered by the regression model is 0.624. The model is well-fitted if this value is more significant than 0.6. The results had an R^2 score of 0.624, higher than 0.6. With an R^2 value of 0.624, the independent variable explains 62.4% of the variation in the dependent variable (WOM). The model is thought to be appropriate.

The F ratio in the ANOVA table, as seen in the above table, determines if the overall regression model matches the data well. The table shows that independent variables statistically affect the dependent variable $F = 45.484$, P 0.05. The regression model fits the data well (see Table 5).

The standardised beta value's significant value, P must be greater than 0.05, showing that the variable significantly and exclusively contributes to the prediction of the dependent variable. If $P > 0.05$, it is considered that the variable does not significantly affect the independent variable's ability to predict outcomes.

Considering the above table, emotional bonds, product quality, self-satisfaction, and self-expressiveness were significant at the confidence level of 0.05. At the same

Table 4 WOM model summary

Model summary					
Model	R	R^2	Adjusted R^2	Std. error of the estimate	Durbin-Watson
1	0.790[a]	0.624	0.610	0.68208	1.869

[a] Predictors: (Constant), IV_Self-expressiveness, IV_Quality of the products, IV_Hygienic factors, IV_Environmental consciousness, IV_Social_status, IV_Financial_benefits, IV_Emotional_bonds, IV_Uncomfortability, IV_Selfsatisfaction
[b] Dependent Variable: DV_WOM

Table 5 ANOVA table WOM

ANOVA[a]						
Model		Sum of squares	df	Mean square	F	Sig.
1	Regression	190.445	9	21.161	45.484	< 0.001[b]
	Residual	114.913	247	0.465		
	Total	305.358	256			

[a] Dependent Variable: DV_WOM
[b] Predictors: (Constant), IV_Self_expressiveness, IV_Quality_of_the_products, IV_Hygienic_factors, IV_Environmental_consciousness, IV_Social_status, IV_Financial_benefits, IV_Emotional_bonds, IV_Uncomfortability, IV_Selfsatisfaction

time, financial benefits, environmental consciousness, social status, hygiene factors, and uncomfortability were insignificant.

The standardised coefficient, Beta 2 for Emotional Bonds, equals 0.227, which indicates that for each unit increase in emotional Bonds, there is an increase in WOM intention of 0.227 times units.

Considering the above Table 6, the unstandardised coefficient, B1 for Financial Benefits, equals − 0.73. This means that for each one unit increase in financial benefits, there is a decrease in WOM intention of 0.0.73 times units.

Regarding product quality and self-satisfaction, the Beta values are 0.119 and 0.520, respectively. When each item increases by 1 unit, the WOM intention will increase by 0.119- and 0.520-times units. The self-expressiveness has a beta value of − 0.103. It guarantees that if self-expression grows by one unit while all other variables remain constant, the WOM intention decreases by − 0.103 units. According to the findings, emotional ties, product quality, and self-satisfaction have favourable effects, but self-expressiveness negatively impacts WOM intention. Therefore, the effect on word-of-mouth intention was shown as significant, and H2 was supported

Table 6 Coefficient WOM

Model		Standardised coefficients	*T*	Sig.	95.0% confidence interval for B		Collinearity statistics	
		Beta			Lower bound	Upper bound	Tolerance	VIF
1	(Constant)		1.121	0.263	− 0.343	1.248		
	IV_financial_benefits	-0.073	− 1.621	0.106	− 0.221	0.021	0.747	1.339
	IV_emotional_bonds	0.227	4.252	< 0.001	0.137	0.372	0.537	1.864
	IV_environmental_consciousness	− 0.006	− 0.127	0.899	− 0.172	0.151	0.716	1.397
	IV_quality_of_the_products	0.119	2.467	0.014	0.033	0.292	0.655	1.526
	IV_selfsatisfaction	0.520	8.037	< 0.001	0.485	0.800	0.364	2.744
	IV_social_status	0.019	0.420	0.675	− 0.092	0.142	0.719	1.392
	IV_hygienic_factors	0.088	1.950	0.052	− 0.001	0.281	0.743	1.347
	IV_uncomfortability	− 0.085	− 1.733	0.084	− 0.243	0.016	0.627	1.595
	IV_self_expressiveness	− 0.103	− 2.278	0.024	− 0.265	− 0.019	0.743	1.347

($0.05 \leq$ p-value), H4 was supported ($0.05 \leq p$-value), H5 was supported ($0.05 \leq P$ value), and H9 was supported ($0.05 \leq p$-value).

15 Summary of the Hypotheses

A summary of the results of the hypotheses testing is reported in Table 7

16 Conclusions and Recommendations

Environmental concerns are emerging as a mainstream issue for modern-day consumers. This critical issue has compelled fashion retailers and brands to embrace sustainable practices that strive to address the environmental concerns of their consumers fully. Given the background, this research has a quantitative approach to provide valuable insights and a better understanding of Sri Lankan fashion consumers' primary factors, concerns, and behaviour concerning the secondhand fashion business.

According to the research, consumers' judgments of value towards secondhand consumption are poorly understood conceptually and practically. There is a significant knowledge gap surrounding secondhand fashion consumption in Sri Lanka. According to the literature review, the study intends to fill a gap in the literature.

Ten hypotheses were developed based on the findings of the qualitative investigation, which were then tested in the quantitative survey study. In terms of purchase intention, only three possibilities are recognised. The acceptance of emotional bonds and self-satisfaction has a positive influence on consumer behaviour, and discomfort has a negative influence on it. Only four hypotheses on the intention of word-of-mouth have been accepted. Consumer behaviour was positively influenced by financial advantages, product quality, self-satisfaction, and acceptance of self-expression, which negatively impacted behaviour. Hence, there is a possibility of consumer trusting their word-of-mouth intention more than their purchasing intention when they are deciding on purchasing secondhand fashion products.

The secondhand fashion business is based on a new business model that must be developed in Sri Lanka. Hence, this research, one of the first studies to investigate and explore the secondhand fashion consumption market amongst Sri Lankan fashion consumers, offers valuable insights into Sri Lankan fashion retailers' sustainability strategies.

The study provides a rich understanding of why fashion marketers must grab this opportunity by considering the marketing of secondhand fashion products to this highly motivated, environmentally conscious consumer segment keen enough to purchase secondhand fashion due to several reasons, such as social and environmental concerns. This research further informs and invites current and future fashion industry practitioners to focus on the key drawbacks and challenges associated with

Table 7 Summary of the hypotheses

Hypotheses	Accepted/rejected
H1 a—financial benefits will positively affect consumers' intention to purchase secondhand fashion products	Rejected
H1 b—financial benefits will positively affect word-of-mouth intention towards Secondhand fashion products	Accepted
H2 a—emotional bonds with the products will positively affect consumers' intention to purchase Secondhand fashion products	Accepted
H2 b—emotional bonds with the products will positively affect word-of-mouth intention towards Secondhand fashion products	Rejected
H3 a—environmental consciousness will have a positive effect on consumer's intention to purchase Secondhand fashion products	Rejected
H3 b—environmental consciousness will positively affect word-of-mouth intention towards Secondhand fashion products	Rejected
H4 a—product quality will have a positive effect on consumer's intention to purchase Secondhand fashion products	Rejected
H4 b—product quality will positively affect word-of-mouth intention towards Secondhand fashion products	Accepted
H5 a—self-satisfaction will positively affect consumers' intention to purchase Secondhand fashion products	Accepted
H5 b—self-satisfaction will positively affect word-of-mouth intention towards Secondhand fashion products	Accepted
H6 a—social status will negatively impact consumer's intention to purchase Secondhand fashion products	Rejected
H6 b—social status negatively impacts word-of-mouth intention towards Secondhand fashion products	Rejected
H7 a—hygienic consciousness will negatively impact consumers' intention to purchase Secondhand fashion products	Rejected
H7 b—hygienic consciousness will negatively impact word-of-mouth intention towards Secondhand fashion products	Rejected
H8 a—uncomfortability will negatively impact consumers' intention to purchase Secondhand fashion products	Accepted
H8 b—uncomfortability will negatively impact word-of-mouth intention towards Secondhand fashion products	Rejected
H9 a—self-expressiveness will negatively impact consumers' intention to purchase Secondhand fashion products	Rejected
H9 b—self-expressiveness will negatively impact word-of-mouth intention towards Secondhand fashion products	Accepted

the secondhand clothing business, mainly raised by non-SHF consumers due to their hygienic and comfortability concerns. The findings of this study reveal the impact on consumers' purchasing behaviour according to their perception values. This clearly explains how consumers behave when purchasing SHF products. SHF retailers can use this as a guide to identify their customers.

16.1 Significance of the Study

There is a vast knowledge gap in Sri Lanka regarding secondhand fashion consumption. Adopting the theoretical perspective of the Means-End-Chain model, this research aims to shed some light on the emerging literature on secondhand consumption by delineating Sri Lankan consumer motivational factors towards secondhand fashion consumption. From a more practical point of view, this research aims to provide insights into Sri Lankan fashion retailers' sustainable strategies, to encourage retailers to focus on consumer values towards secondhand fashion consumption.

16.2 Limitations and Future Research Direction

This research is subject to a few limitations, which will likely offer scope for further research. First, the current study is contained by quantitative research approaches, which gathering huge data sets it may limit the research findings' applicability and analysis timeframe. Since it is beyond the scope of this study, further research should be undertaken as a cross-cultural/variant study to acquire a more comprehensive understanding of consumers' views of value when it comes to secondhand fashion consumption. Future research should also investigate whether the communication and marketing tactics employed by secondhand retailers are adequate and whether they have an impact on the adoption of secondhand fashion product purchases. It could also be interesting to collect data from various parts of the country and combine it with information about people's educational backgrounds, income levels, age, gender, and place of origin. Even though the information about genuine establishments that provide secondhand fashion products is limited, it should be evaluated. Buying behaviour is sometimes influenced by the firm's marketing techniques rather than the individual's initiative to seek out these things. As a result, future research on impulse purchases may be worthwhile.

In the current economic crisis in Sri Lanka, there is a huge opportunity for a secondhand fashion business model. Since it will be a great support to the community as well as their purchasing decision during the economic crisis. Future studies, on the other hand, may be directed at consumer behaviour in the era of the Covid-19 pandemic. It might be useful to examine their behaviour throughout the epidemic and come up with viable solutions to the limits that secondhand businesses confront.

This work was supported by the Senate Research Council, University of Moratuwa, under (SRC/LT/2021/11).

References

1. Allwood JM, DeRodriguez MC, Bocken NM, Bocken NM (2006) Well dressed? The present and future sustainability of clothing and textiles in the United Kingdom [online]. Available at http://www.ifm.eng.cam.ac.uk/resources/sustainability/welldressed/. Accessed on 14 May 2014
2. Bocken NMP, de Pauw I, Bakker C, van der Grinten B (2016) Product design & business model strategies for a CE. J Industr Prod Eng. https://doi.org/10.1080/21681015.2016.1172124
3. Bardhi F, Arnould EJ (2005) Thrift shopping: combining utilitarian thrift and hedonic treat benefits. J Consum Behav 04:223–233
4. McKinsey (2020). McKinsey & Company launched global banking annual review 2020. The sustainable clothing action plan 2020. WRAP. https://wrap.org.uk/taking-action/textiles/initiatives/scap-2020
5. Lehmann M, Arici G, Boger S, Martinez-Pardo C, Krueger F, Schneider M, Carrierepradal B, Schou D (2019) Pulse of the fashion industry
6. Choi TM, Cheng TE (2015) Sustainable fashion supply chain management. Springer, New York
7. Shim S (1995) Environmentalism and consumers' clothing disposal patterns: an exploratory study. Cloth Text Res J 13(1):38
8. Morgan M-J (2020) Plastics and the environment; these facts show how unsustainable the fashion industry is. https://www.weforum.org/agenda/2020/01/fashion-industry-carbon-unsustainable-environment-pollution/
9. Rukhaya S, Yadav S, Rose NM (2021) Arpita grover and diksha bisht. A sustainable approach to counter the environmental impact of fast fashion. Pharm Innov J 10(8S):517–523
10. Morgan LR, Birtwistle G (2009) An investigation of young fashion consumers' disposal habits. Int J Consum Stud 33(2):190–198
11. Larsson J, Mouwitz P, Peterson J (2009) Knit on demand: mass customisation of knitted fashion products. Nordic Text J
12. Tammemagi HY (1999) The waste crisis: landfills, incinerators, and the search for a sustainable future. Oxford University Press
13. Geegamage T, Ranaweera A, Halwatura R (2021) Secondhand fashion consumption: a literature review. Paper presented at the ERU symposium 2021
14. Council for Textile Recycling (2014) Abingdon. http://www.weardonaterecycle.org/about/issue.html. Accessed on 17 Feb 2014
15. Maegan ZW, Yan R (2013) An exploratory study of the decision processes of fast versus slow fashion consumers. J Fash Mark Manag 17(2):141–159
16. Henninger CE (2015) Traceability the new eco-label in the slow-fashion industry? Consumer perceptions and micro-organisations responses. Sustainability 7(5):6011–6032
17. To E, Kan K (2020) China's next problem is recycling 26 million tons of discarded clothes the world's biggest fashion buyer is running out of room for all its used garments. Bloomberg Green, News
18. Allwood JM, Laursen S, Russell S, Rodríguez C, Bocken N (2008) An approach to scenario analysis of the sustainability of an industrial sector applied to clothing and textiles in the UK. J Clean Prod 16:1234–1246. https://doi.org/10.1016/j.jclepro.2007.06.014
19. Shirvanimoghaddam K, Motamed B, Ramakrishna S, Naebe M (2020) Death by waste: fashion and textile circular economy case. Sci Total Environ 718:137317. https://doi.org/10.1016/j.scitotenv.2020.137317
20. Domina T, Koch K (1999) Consumer reuse and recycling of post-consumer textile waste.

21. Granzin KL, Olsen JE (1991) Characterising participants in activities protecting the environment: a focus on donating, recycling, and conservation behaviors. J Public Policy Mark 10:1–27. https://doi.org/10.1177/074391569101000201
22. Xu Y, Chen Y, Zhao RH (2014) Secondhand clothing consumption: a cross-cultural comparison between American and Chinese young consumers. Int J Consum Stud 38(06):670–677
23. Joung HM (2014) Fast-fashion consumers' post-purchase behaviours. Int J Retail Distrib Manag 42(8):688–697
24. Iran S, Schrader U (2017) Collaborative fashion consumption and its environmental effects. J Fash Mark Manag 21:468–482. https://doi.org/10.1108/JFMM-09-2016-0086
25. Guiot D, Roux D (2010) A secondhand shoppers' motivation scale: antecedents, consequences, and implications for retailers. J Retail 86(4):383–399
26. Gopalakrishnan S, Matthews D (2018) Consumer attitudes and communication in circular fashion. J Fash Mark Manag 22(3):189–208
27. ThredUp (2019) Thredup 2019 resale report, thredup. Available at https://www.thredup.com/resale/2019?tswc_redir5true. Accessed 10 Aug 2020
28. Bankole A, Abdurrahaman M, Oyeranti O (2004) Cotton and textile production and trade of small producers and their livelihoods in the cotton belt of nigeria: a contextual analysis within the framework of fair trade. Oxfam GB, Abuja
29. Zanjirani Farahani R, Asgari N, Van Wassenhove LN (2022) Fast fashion, charities, and the circular economy: challenges for operations management. Prod Oper Manag 31:1089–1114. https://doi.org/10.1111/poms.13596
30. Albinsson PA, Perera BY (2009) From trash to treasure and beyond the meaning of voluntary disposition. J Consum Behav 08(06):340–353
31. Vinson DE, Scott J, Lamont L (1977a) The role of personal values in marketing and consumer behavior. J Market 41:44–50; Management 22(3):286–300
32. Vinson DE, Munson JM, Nakanishi M (1977b) An investigation of the rokeach value survey for consumer research applications. In: Perreault W (ed) Advances in consumer research. Association for Consumer Research, Chicago, IL, No. 14, pp 247–252
33. Rokeach M (1973) The nature of human values. The Free Press, New York, NY
34. Scott JE, Lamont LM (1977) Relating consumer values to consumer behavior: a model and method for investigation. In: Greer TW (ed) Increasing marketing productivity. American Marketing Association, Chicago, IL, pp 283–288
35. Jackson T (2005) Motivation sustainable consumption. A review of evidence on consumer behaviour and behavioral change. Report to the DEFRA sustainable development research network. Retrieved
36. Hines JD, O'neal GS (1995) Underlying determinants of clothing quality: the consumers' perspective. Cloth Text
37. Gutman J (1984) Analyzing consumer orientations toward beverages through means–end chain analysis. Psychol Market 1(3–4):23–43
38. Overby JW, Woodruff RB, Gardial SF (2005) The influence of culture upon consumers' desired value perceptions: a research agenda. Mark Theor 5:139–163. https://doi.org/10.1177/1470593105052468
39. Reynolds TJ, Gutman J (1984) Advertising is image management. J Adv Res
40. Zachariah ZB, Jusan MBM (2011) Means-end chain model framework for measuring housing environment choice behavior. J Civ Eng Archit
41. Geegamage T, Ranaweera A, Halwatura R (2023) Pre-loved or hatred? Consumers' perception of value towards secondhand fashion consumption in Sri Lanka. Res J Textile Apparel
42. Hur E (2020) Rebirth fashion: secondhand clothing consumption values and perceived risks. J Clean Prod 273:122951. https://doi.org/10.1016/j.jclepro.2020.122951
43. Yu S, Lee J (2019) The effects of consumers' perceived values on intention to purchase upcycled products
44. Laitala K, Klepp IG (2018) Motivations for and against secondhand clothing acquisition. Cloth Cult 5(2):247–262

45. Park H, Kwon TA, Zaman MM, Song SY (2020) Thrift shopping for clothes: to treat self or others? J Glob Fash Market 11:56–70
46. Kim I, Jung HJ, Lee YJS (2021) Consumers' value and risk perceptions of circular fashion: comparison between secondhand. Upcycl Recycl Cloth 13:1208
47. Yoo F, Jung HJ, Oh KW (2021) Motivators and barriers for buying intention of upcycled fashion products in China. Sustainability 13:2584. https://doi.org/10.3390/su13052584
48. Joung HM, Park-Poaps H (2013) Factors motivating and influencing clothing disposal behaviors. Int J Consum Stud 37:105–111
49. Machado MAD, Almeida SOD, Bollick LC, Bragagnolo G (2019) Secondhand fashion market: consumer role in the circular economy. J Fash Market Manage 23(3):382–395. https://doi.org/10.1108/JFMM-07-2018-0099
50. Turunen LLM, Leipämaa-Leskinen H (2015) Pre-loved luxury: identifying the meaning of secondhand luxury possessions. J Prod Brand Manage 24(1):57–64
51. Cervellon MC, Carey L, Harms T (2012) Something old, something used: determinants of women's purchase of vintage fashion vs secondhand fashion. Int J Retail Distrib Manage 40(12):956–974
52. Thompson SCG, Barton MA (1994) Ecocentric and anthropocentric attitudes toward the environment. J Environ Psychol 14:149–157
53. Vehmas K, Raudaskoski A, Heikkila P, Harlin A, Mensonen A (2018) Consumer attitudes and communication in a circular fashion. J Fashion Market
54. Edbring EG, Lehner M, Mont O (2016) Exploring consumer attitudes to alternative models of consumption: motivations and barriers. J Clean Prod 123:5–15
55. Silva SC, Santos A, Duarte P, Vlačić B (2021) The role of social embarrassment, sustainability, familiarity and perception of hygiene in secondhand clothing purchase experience. Int J Retail Distrib Manage 49(6):717–734. https://doi.org/10.1108/IJRDM-09-2020-0356
56. Park HS, Kim HS (1998) A study on the dimensions of clothing attitudes. J Korean Soc Cloth Text 22:279–289
57. Kim CJ (1991) Transactions: risk perceived by consumers in apparel buying situation (I)—risk types and their relationships with consumers' demographic variables. J Korean Soc Cloth Text 15:405–416
58. Creswell JW (2003) A framework for design. Research design: qualitative, quantitative, and mixed methods approaches, 9–11
59. Jaccard J, Helbig DW, Wan CK, Gutman MA, Kritz-Silverstein DC (1990) Individual differences in attitude-behavior consistency: the prediction of contraceptive behavior. J Appl Soc Psychol 20(7):575–617

The Use of ISO 14062:2002 Sustainable Design Strategies of the Designers to Address SDG 12: Ensure Sustainable Consumption and Production Patterns in the Export Apparel Industry in Sri Lanka

Prabod Dharshana Munasinghe

Abstract The textile and apparel industry is the second largest polluter in the world; hence, it is the primary export industry in Sri Lanka. Because of this high negative impact level of the industry, this chapter explains a study that investigated the impact of the export apparel industry in Sri Lanka on Sustainable Development Goal (SDG)12: ensure sustainable consumption and production patterns to concise possible regulatory guidelines and suggestions for the Sri Lankan apparel Industry. Also, ISO 14062:2002 standards introduced eight sustainable product design strategies that can be used to ensure the goals of SDG12. Sri Lankan Apparel exports mainly cater to mass-market retailers; therefore, this study was conducted with mass-market-oriented apparel exports to get a broader impact. The study employed a mixed method where qualitative and quantitative elements were examined. Thirty-five designers with 3–20 years of experience in this sector were questioned through an online questionnaire. The answers were analysed based on a quantitative approach to find patterns. Ten designers with more than ten years of experience were interviewed to clarify the generated patterns of the online questionnaire. The outcomes revealed that the design process in mass-market-oriented apparel exports is mainly based on profits. However, they have to pay deep concentration to sustainability due to the pressure from external policies such as governments and other non-government organisations. Furthermore, the export apparel designers in Sri Lanka use ISO 14062:2002 sustainable design strategies based on two perspectives: profit-oriented; and policy-oriented. Specific strategies still need to be considered thoroughly while designing with profit orientation. Also, the effective use of land is heavily neglected during the design process due to the lack of awareness. However, it can make an impact on both profit and policy orientations. Based on the findings, the study suggested that the export apparel designers need more awareness, training, and support from higher management to implement the design strategies. Also, it is found that ISO

P. D. Munasinghe (✉)
Department of Textile and Apparel Engineering, University of Moratuwa, Moratuwa, Sri Lanka
e-mail: prabodm@uom.lk

G. M. Ranathunga (ed.), *Textiles of Sri Lanka*, SDGs and Textiles,
https://doi.org/10.1007/978-981-97-0569-6_7

14062:2002 related sustainable design strategies should be refined and eloborated to get more outcomes to achieve SDG12. Finally, interpreting sustainable design strategies from a profit-oriented perspective can attract the mass-market apparel industry to actual, sustainable initiatives to achieve SDG12 more realistically. Therefore, more explorations are needed to fulfil that requirement.

Keywords SDG 12: ensure sustainable consumption and production patterns · ISO 14062:2002 · Sustainable design strategies · Export apparel industry · Sri Lanka

1 Introduction

Sustainable development is a keyword that emerged in the 1970s [1] and was later brought to a comprehensive level in 1987 by the United Nations (UN) through the Brutland Report [2]. The main reason to establish this topic was that they had noticed the heavy risk created by rapid industry developments without compromising the world's future. The UN has taken several measures to balance the world's sustainability; however, those attempts were merely supported to obtain a sensible impact. Therefore, they have seriously discussed this matter and introduced Sustainable Development Goals (SDGs) in 2015 with the member countries' agreement to be achieved in 2030 [3]. Sri Lanka is also a UN member committed to achieving the SDGs. There are 17 SDGs, and this chapter mainly focuses on SDG12: Ensure sustainable consumption and production patterns [4]; hence, to ensure sustainable consumption and production, the product development and manufacturing processes play a vital role [5].

According to the most updated statistics, the textile and apparel industry is the second largest polluter in the world, which challenges SDG12. China is the world's largest apparel exporter, accounting for US$176 billion worth of business, while Sri Lanka is one of the key export partners, accounting for US$5.9 billion in business, which consists of ethically and environmentally high standards. Even though it takes a tiny portion of the global apparel export business, it is the main export revenue generator of the country Sri Lanka, which heavily depends on Apparel exports now. On the other hand, Sri Lanka is a 65,610 km^2 small Island with nearly 20 million population. The population density in 2020 was 354 people per square kilometre. It was 264 people per square kilometre in 1987, which shows that the growth is around 34% in the last 35 years [6]. From 1990 to 2020, Nitrogen Oxide emission in the country increased by 57%, and Carbon dioxide emission from liquid fuel consumption increased by 349% [7]. With the limited space on the Island and the incremental behaviour of the pollution indexes, it is vital to investigate critical stakeholders in the country's manufacturing capacity to find the current trends and plan future directions related to the SDS12.

The Sri Lankan textile and apparel industry is the largest exporter, which accounts for 52% of the total local exports. The size of the apparel export market is amounting to US$5.9bn in 2022. The United States, the United Kingdom, and Europe are the major partners in the industry [8]. Private entities entirely manage the industry. Therefore, government-level institutions' active and complete involvement in addressing the SDGs still needs to be improved. However, the industry maintains its best standards to attract buyers, such as zero child labour, women empowerment [9], and specific environmental improvements such as renewable resource use and minimising water pollution. Sustainable production has a significant focus. However, the involvement to improve the link between production and consumption is yet to be improved [10].

The structure of the Sri Lankan export apparel industry consists of manufacturing-based companies; however, they offer value-adding services such as design, product development, and innovations [11]. The sustainability improvements should be considered during the early stages of the design process rather than once the product is made [12]. The product design, development, and innovation-related activities can firmly decide the final product characteristics, and the front-end teams, such as design, product development, and innovations, can influence the product [13, 14].

There are many design strategies to drive the front-end teams [15–17]. However, international standards should be considered to produce a global-level recognised product. Therefore, ISO has introduced a set of sustainable design approaches such as improvement of material efficiency, improvement of energy efficiency, sparing use of land, design for cleaner production and use, design for durability, design for optimising functionality, design for reuse recovery and recycling, and avoid the hazardous substances [15]. However, whether the front-end teams use these strategies in their day-to-day activities is still being determined.

Therefore, this chapter aims to explore the existing use of ISO 14062:2002 sustainable design approaches of the design teams in Sri Lankan apparel exporters to offer recommendations to reshape the existing practices to ensure the SDG12 in the Sri Lankan export apparel Industry.

2 Literature Review

This section comprehends the context and rationale for the study.

2.1 Sustainable Development Goals

Sustainability with environmental balance was initially discussed by the United Nations in 1972 in Stockholm, and they have identified that the existing practices would have a severe negative impact on the planet in the 2020s. At that time, they predicted that the rise of human needs and aspirations would reduce the world's resources and peace [18] (Fig. 1).

When evaluating this forecast of the UN, it is justified that the prediction of the UN is highly accurate. UN estimated that every year, the freshwater demand has

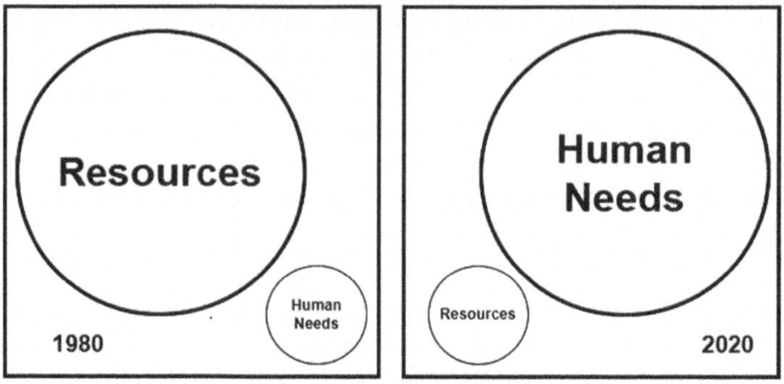

Fig. 1 UN projection in 1972 regarding resource availability versus human needs on the planet

increased by 1% from 1980 onwards [19]. The demand for freshwater, minerals, and land has rapidly increased in the last two decades.

Therefore, the UN introduced several strategies to address this issue, and the latest policy-level framework is Sustainable Development Goals, which were introduced in 2015. These SDGs can be classified into three pillars: Economic, Environmental, and Social. The SGSs such as 1, 2, 3, 8, and 9 are categorised under the economic pillar, 6, 7, 12, 13, 14, and 15 are categorised under the environment pillar, and 4, 5, 10, 11, 16, and 17 are categorised under the social pillar [20]. Out of all the three pillars, most of the main discussions are interwoven in this decade around the environment pillar because the timeline of the SDGs is nearly halfway completed; the outcomes still need to be satisfying the required levels [21].

2.2 *SDG 12: Ensure Sustainable Consumption and Production Patterns*

According to Kostoska and Kocarev [20], the environmental pillar consists of six SDGs and SDG 12: Ensuring sustainable consumption and production patterns plays a vital role among them. SDG 12 mainly focuses on balancing economic and environmental aspects; for example, it describes the need to "do more with less" [4]. The modern world is rapidly growing with numerous economic models; therefore, the needs of humans are also rapidly widened [22]. As a result of the changes in human needs, consumption and production patterns have also rapidly changed in the last two decades.

For example, Charles Frederick Worth invented the modern fashion system at the end of the nineteenth century, and the mass production of apparel was introduced as a ready-to-wear (RTW) concept. At the end of the twentieth century, the different market levels and retailing models were developed based on this RTW concept [23, 24]. However, the fashion business was interweaving as a seasonal business, introducing new collections each season. Mainly, there were four seasons. After the first decade of the twenty-first century, brands like H&M and Zara restructured that fifty-year-old marketing methodology. They shortened the lead times of the seasonal business from 12 to 01 month by introducing new collections each month [25]. Also, there was a boom in social media usage, and due to it, peoples' needs were also rapidly changed. As a result of this moment, ZARA further reduced its lead times; therefore, now they released collections each week. This new model rapidly changed the buying behaviours of apparel consumers; they wanted to frequently change their looks to fulfil particular social media needs and other day-to-day activities. Frequent buying draws retailers to introduce low-priced bulk productions, ultimately creating a huge environmental burden [26]. This example clearly shows the changes in modern human needs supported a lot to change the consumption and production patterns. Ultimately, those changes challenged the planet, especially the stress on virgin and non-renewable resources and pollution. As a result of these consumption and production patterns, the apparel industry is now identified as the second largest environmental polluter in the world [27].

Another key example of the importance of SDG 12 is the way that the environment behaves during the pandemic. Due to the COVID-19 pandemic, all activities in the world were paused for more than six months; because of it, the quality level of the air in India was rapidly improved. Generally, the air quality in India's leading cities is deficient due to the emissions in transportation and industrial environments [28]. However, the pandemic paused most transportation and manufacturing activities, lowering emissions. As a result, the air quality has rapidly improved [28], and India and many other Asian countries, including Sri Lanka, which has high air emissions, benefited from the same movement [29]. The pandemic has reduced consumption and production behaviours, which ultimately recover the quality of the environment, which highlights the importance of taking robust actions focusing on SDG12.

The consumption and production patterns in the apparel Industry also play a vital role in supporting SDG12. For example, the Bangladeshi apparel industry emits unbearable waste to the water streams due to the wet processing. Even though the government established laws to establish effluent treatment plants, most companies are reluctant. As a result of it, the quality of the water became very low, and the people's livelihood became a challenge [30]. Nearly 15%–18% of waste generation is evident when creating 1 kg of garments in the Sri Lankan Apparel Industry [31], and washing and dying factories significantly contribute to water pollution [32]. These impacts could be unbearable for a small island like Sri Lanka.

Sri Lanka is a crucial apparel exporter, which contributes as the leading foreign exchange earner of the country. Therefore, it is vital to investigate its contribution to sustainable consumption and production patterns. Also, previous scholars emphasised that sustainable consumption and production patterns can be regulated during the early stages of a product design process [33], and it is challenging and costly when considering sustainability once the product is being made [12]. Furthermore, it is understood that the product design process decides the final product characteristics, and the product designer influences the product features [13].

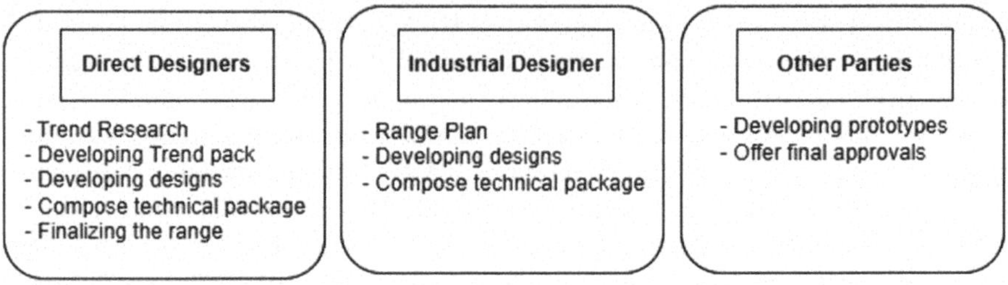

Fig. 2 Role of the direct designer and the industrial designer [11]

2.3 Designer's Role in the Export Apparel Industry in Sri Lanka

When investigating the export apparel statistics of Sri Lanka, it is evident that the primary market share belongs to the mass-market apparel retailers such as Amazon, ASOS, H&M, … etc. [34]. The main manufacturing capacities were allocated to this market segment. Major apparel manufacturers such as MAS Holdings, Brnadix Apparel Limited, and Hiradaramani International Exports Private Limited play a vital role in these manufacturing processes [35]. All these manufacturers have their design teams that directly deal with the retailers. When considering the design process of the mass market, there are two types of design teams: direct designers who work in the design teams of the retailers as well as the industrial designers who work for manufacturers like MAS and Brandix who play a supportive role in adding value to the product development process [11] (Fig. 2). The design teams of these local manufacturers play that industrial designer role. They do their designs based on the directions of the direct designers, as well as involvement in the product development process of the designs of direct designers. These industrial designers are involved in the decision-making process of the product as well as its sustainability [36].

2.4 Sustainable Design Strategies in ISO 14062:2002 Standards

Since the designer has more power to regulate the product in the design stage [12] and the export apparel industry-related designers in Sri Lanka have the same strength to influence the global retailers through their design process, it is vital to investigate the standard procedures to incorporate sustainable elements into the designs [36]. Therefore, the standards of the International Standard Organisation play a vital role. International Standard Organisation introduced eight design strategies to be used in the design process in 2002 [15]. They are,

Strategy 01: Improvement of Material Efficiency

The material is the crucial element of a product, which should have high efficiency to get the relevant outcome. From the apparel perspective, the primary materials are the fibres. Certain materials are also used for trims, value additions, … etc. [37]. For example, traditional cotton has high absorbency; however, new fibre innovations perform more than general cotton [38]. Spandex is one of the common materials that is used to get high stretchability. However, its environmental performance is deficient [39]. Latex is one of the alternatives for stretchable solutions, which has better biodegradability than spandex and artificial rubber [40]. Therefore, designers should have a clear idea about the effectiveness of each material in order to make robust decisions.

Strategy 02: Improvement of Energy Efficiency

Also, the designer should be aware of improving the product's energy efficiency while designing it [41]. Specific manufacturing methods, machinery and tools utilise higher amounts of energy. Also, specific energy sources are nonrenewable [41]. Therefore, the designer should be aware of renewable energy sources and highly efficient production methodologies. Also, they should focus on the user phase since it consumes very high energy [42]. Therefore, designers can make intelligent design decisions to incorporate advanced technologies to reduce energy usage in the user phase.

Strategy 03: Sparing Use of Land

The textile and apparel industry utilises 5% of the cultivated lands globally, which will be increased to 7% in 2030 [43]. Material manufacturing uses much land, while other processes also need a certain amount for their functions [44]. Specifically, natural fibre production consumes high land usage [12]. One of the critical decisions of the design process is material selection, which has a direct link with land usage. Therefore, education on land usage can make a significant impact on the sustainability of the industry.

Strategy 04: Design for Cleaner Production and Use

Cleaner production and cleaner use are vital themes in the twenty-first century, and designers have substantial power to regulate those two areas [45]. Design decisions can support cleaner manufacturing approaches. For example, renewable sources support cleaner production; the designer can decide it in the design process. Also, the designer's decision related to the user's activities helps keep the cleaner use. For example, if a designer chooses a wrinkle-free material, it saves energy; if the designer chooses a low temperature for washing, it supports cleaner use in the use phase of the product [42, 46]. Therefore, designers can strongly impact cleaner production and use through the design process.

Strategy 05: Design for Durability

Durable designs support sustainability in many ways. Firstly, the durable design can avoid new purchases. Secondly, a durable design extends the life span of a product, which stretches the timeline between the use phase and the landfilling stage and reduces the amount of landfilling [47, 48]. Also, durable designs support reuse and upcycling concepts, removing the product from its grave [47]. Therefore, the designer's involvement in improving the product's durability can strongly impact its sustainability.

Strategy 06: Design for Optimising Functionality

Optimising functional values also supports extending the product life while avoiding new purchases. For example, a multifunctional garment can cater for several functionalities and merge several garments into a single design that removes the user from multiple purchases [49, 50]. Avoiding new purchases can save the use of virgin materials, millions of energy and other resources. Furthermore, this strategy supports the reuse and upcycling abilities, extending the product's life.

Strategy 07: Design for Reuse Recovery and Recycling

Regulating the post-manufacturing stage is vital for a sustainable future. Therefore, the designer should regulates the product's sustainability in the post-manufacturing stages and designers need to pay more attention to strengthening the sustainability potential of a product after its use. However, if the designer can incorporate design features for reuse [51], recycling [52], and recovery [52] capabilities, the product will move out from landfilling. Reuse can widely extend the product life while recovery and recycling can support closing the loop [53]. Closing the loop of the apparel product lifecycle can save millions of virgin raw materials, including non-renewable sources like crude oil and thirsty crops like cotton [54].

Strategy 08: Avoid Hazardous Substances

The apparel industry also uses millions of chemicals and emits millions of hazardous substances to the environment. The wet processing and other value additions like printing support this negative impact [55]. Over 8,000 synthetic chemicals are used in the apparel industry, and carcinogenic and hormone-disrupting substances can be found among these chemicals. Furthermore, flame retardants, AZO dyes, chromium, and formaldehyde can also be found in that list [56]. Therefore, improving designers' awareness of hazardous substances can avoid using hazardous elements before making the product.

The above descriptions emphasise the importance of having strong awareness and readiness among designers to use the ISO 14062:2002 design strategies to make a truly sustainable industry.

3 Methodology

A mixed approach was used to collect the data. Initially, an online-based survey was circulated to the design teams located in the export-oriented market. Purposive and snowball sampling methods were used to circulate the survey. First, a managerial-level person in those design teams was contacted, and the other design team members were contacted through their channels. It was asked to circulate the questionnaire across all the levels of the design team to get a broad idea about the awareness in different experience and responsibility levels. Thirty-five designers with more than three years of experience participated in the survey. Based on the questionnaire outcomes, selected ten designers were again contacted and conducted discussions to clarify the outcomes. The interview guide was used in this phase.

3.1 Online Survey

The online survey was developed to comprehensively understand the awareness and usage of the ISO 14062:2002 standards-related design strategies. This standard has eight strategies, and those eight components were further dismantled to offer easy understanding to the designers, as given in Table 1.

Designers were asked to evaluate their use of these strategies based on the given 11 areas and rate them under four levels: always consider when designing, sometimes

Table 1 Covered areas in the online questionnaire

#	Standard area in ISO 14062:2002 [13]	Subtopics used in the questionnaire	Question number
1	Improvement of material efficiency	Improvement of material efficiency of the product design	01
2	Improvement of energy efficiency	Improvement of energy efficiency of the production through the product design	02
3	Sparing use of land	Economical use of land through the product design	03
4	Design for cleaner production and use	Design for cleaner production	04
		Design for cleaner use of the product	05
5	Design for durability	Design to improve the durability of the product	06
6	Design for optimising functionality	Design for optimising functionality	07
7	Design for reuse recovery and recycling	Design for reuse	08
		Design for recovery	09
		Design for recycling	10
8	Avoid hazardous substances	Avoid hazardous substances	11

Table 2 Scoring system

Answer	Maximum score
Always consider this when designing	3
Sometimes consider	2
Have no opportunity to consider	1
Do not know this area	0
Total questions	11
The maximum obtainable total score of a designer due to best practices	$3 \times 11 = 33$

consider, have no opportunity to consider, and do not know this area. Those answers were quantified through a scoring system by adding a value to each answer, as in Table 2. There were 11 questions, and the maximum score obtained for a designer was 33.

3.2 *Semi-Structured Interviews*

Based on the survey results, specific patterns and identical features were identified for further investigation to get a clear idea. Ten designers with more than ten years of experience in the field were contacted again and conducted 15-min discussions via phone calls.

The deductive approach was used to analyse the data because the themes of the investigation were predefined based on ISO 14062:2002 standards [13]. Based on those themes, the principal analysis was conducted. The scores mentioned above were considered to make the interpretations and the analysis.

4 Results

The questionnaire was distributed to 35 designers who are working in the export apparel industry in Sri Lanka. The sample consists of the designers working on export in the industry. The experience of the sample was varied from 3 to 20 years (Fig. 3). Interviews were conducted with designers with more than ten years of experience to get more in-depth insights.

4.1 *Improvement of Material Efficiency*

Material efficiency is one of the critical topics discussed in the industry, and most material manufacturers bring eco-friendly materials to the context. Therefore, the

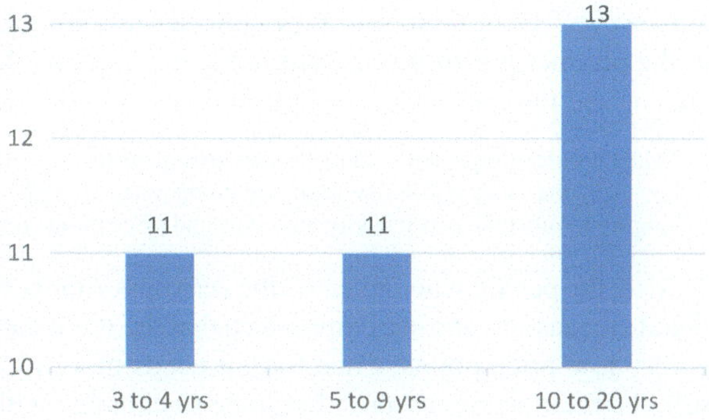

Fig. 3 Experience distribution across the sample

awareness of sustainable materials was very high among the sample; hence, all the respondents had good awareness. However, using this design strategy in their day-to-day work is vital. Twenty-two said they always use this strategy in their design process, while 12 used it in specific applications. One of the designers emphasised that he had no freedom to use this strategy in his daily work schedules (Fig. 4).

These variations are deeply questioned during the interviews, and it is revealed that the trend towards sustainable approaches in most of the mainstream apparel brands pressures the designers to use sustainable materials in their products. Therefore, the majority of them consider this strategy in the design process. Most apparel brands catering to the high street have set goals to convert their entire product portfolio into a high level of sustainability using eco-friendly materials. Certain brands claim that they will use 50% sustainable materials in 2025, while others claim that they will use 100% sustainable materials in 2030. These approaches pressure the designers in Sri Lanka to consider environmentally efficient materials.

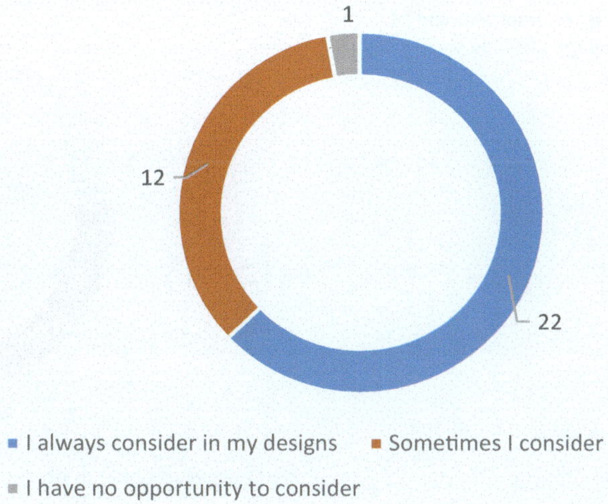

Fig. 4 Improvements in material efficiency

However, 12 of them said they consider this strategy in some instances because all the product portfolios are still not using sustainable materials. One interviewee emphasised the cost factor and the challenge of considering eco-friendly materials.

> "Most retailers request to include sustainable materials while keeping the same price points; however, the sustainable materials are comparatively high in price. Therefore, it is challenging to meet the eco-friendly materials and price points related goals simultaneously…."

Also, the person who indicated the zero opportunity to use this strategy questioned to find the reasons, and she emphasised that she is allocated to a product range catering to very low-price points. Therefore, it needs more room to consider sustainability. Still, the retailers do not push her to be sustainable with that specific product line.

This evidence shows that the industry is severely hungry to convert material perspectives into eco-friendly concepts. However, the existing business models in the mass market do not offer freedom to achieve the goals.

4.2 *Improvement of Energy Efficiency*

The results reveal that improving the product's energy efficiency is challenging for designers in the mass market. Only 9 (26%) said they are considering reducing energy usage through the product design. 50% of them said they can only consider this in certain circumstances. At the same time, 20% have no opportunity to consider it. The exciting finding is that a person with five years of experience in the field has yet to learn about this strategy (Fig. 5).

These results revealed the fragmented view of designers regarding saving energy of the product lifecycle. Therefore, it was questioned during the interviews as well.

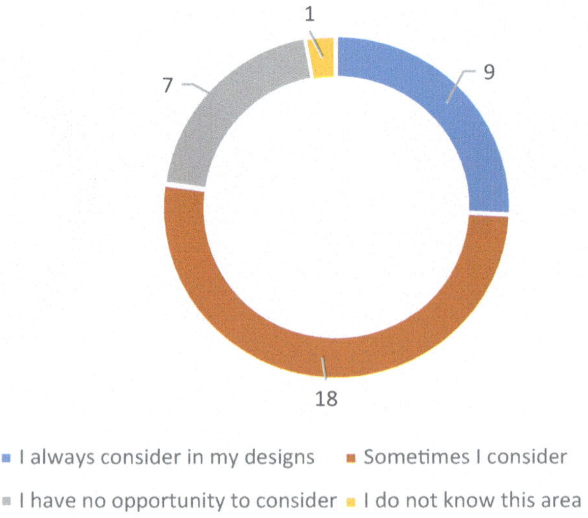

Fig. 5 Improvement of energy efficiency

Export manufacturing plants in Sri Lanka heavily use solar power. However, the designer needs more information regarding those approaches to make robust decisions to save energy. On the other hand, material manufacturing related energy information is not readily available, and the designers need to gain more knowledge to use that information in the design process.

4.3 Sparing Use of Land

Considering land use while optimising the sustainability of a product during the design process takes work. The designers should have comprehensive knowledge about this factor to practically use in the design process. Figure 6 clearly shows that the majority are not considering it in the design process, while three have yet to learn about this strategy, and they have 13, 10, and 5 years of experience in the field. Those people were questioned to find the reason for their lack of awareness, and then it was revealed that the majority of the sample thinks sparing use of land relates to space usage in their manufacturing plants. None of the ten interviewees had a good idea about different materials that use less land in fibre production. An interviewee with 11 years of experience said, "I have never thought about this angel in my career." Another designer with 18 years of experience said "we rarely get such information from the suppliers, and no one inspired us to think about land use in this view earlier…".

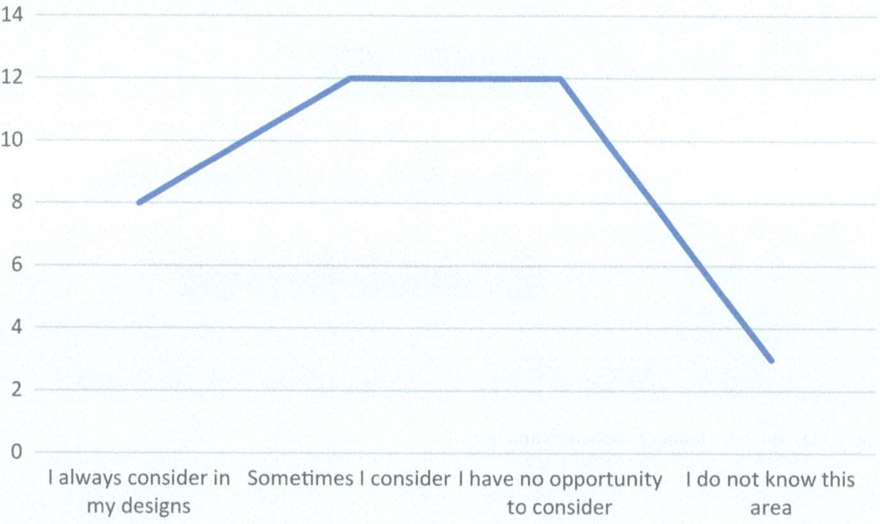

Fig. 6 Sparing use of land

4.4 Design for Cleaner Production and Use

This strategy was questioned from two angles, as mentioned in Table 2, questions 04 and 05: product manufacturing and use perspectives. The results reveal that designers have similar awareness and freedom in using this strategy. Three of them have yet to learn about the term cleaner production, and one of them has 13 years of experience in the industry. Cleaner production is one of the critical areas that the fashion world should thoroughly consider. However, Fig. 7 shows that most designers lack the freedom to consider it in their daily routines.

Also, it is revealed during the interviews that retailers and manufacturers consider improving their processes to make a cleaner production. However, they need more motivation from retailers and the company owners that they are working, to consider cleaner use. Lack of motivation in the current context can occur due to two reasons: mass-market retailers are pressurised by the governments to make their products more sustainable, which urges retailers to pay more consideration to their production processes than use phase, or retailers and manufacturing companies have a lack of awareness about the importance of considering the use phase of the product in terms of sustainability. However, the use phase strongly impacts the imbalance of overall environmental sustainability of a product. Therefore, the designers should have a robust motivation to consider this factor in their design process.

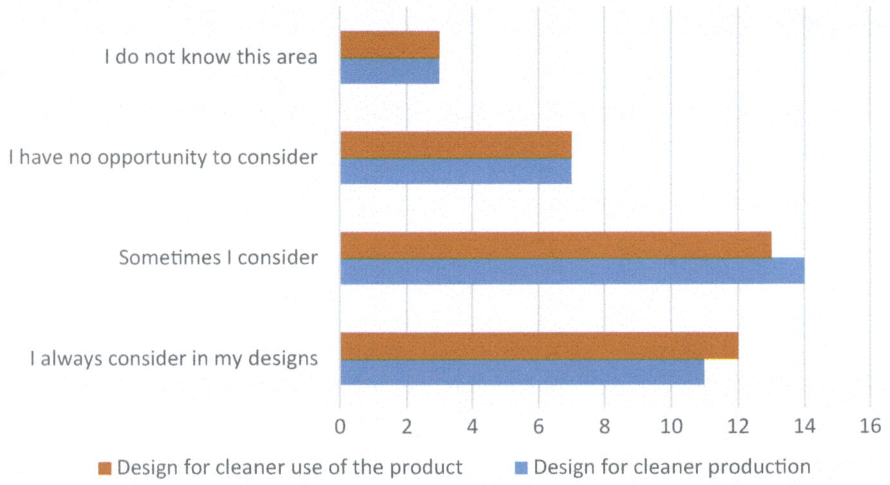

Fig. 7 Design for cleaner production and use

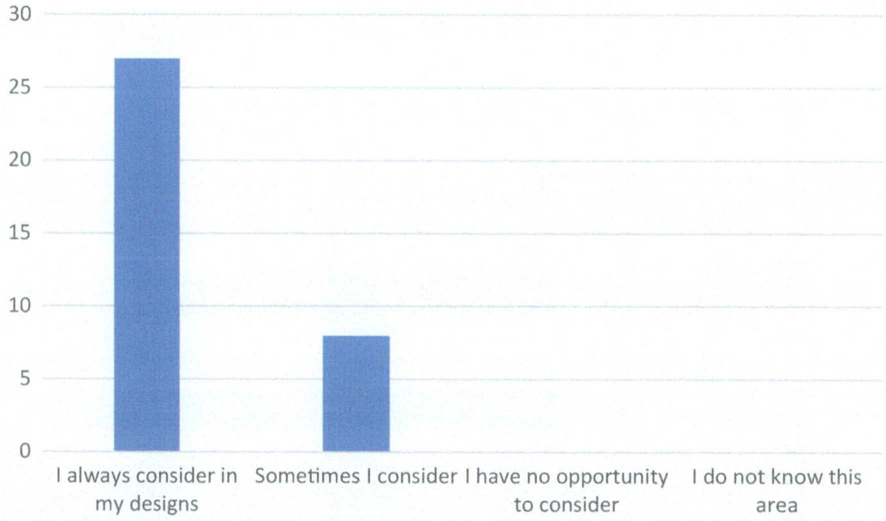

Fig. 8 Design for durability

4.5 Design for Durability

Durable design is one of the critical needs of a customer. Every customer expects long-lasting products. That requirement can be seen in the results of the study as well. Twenty-seven (77%) say they always consider the product's durability in the design process, while 23% say they consider it on certain occasions. However, they still consider this strategy in their work (Fig. 8). Four designers who are considering this strategy at certain times, have more than ten years of experience. Therefore, it was questioned during the interviews. The interviews revealed that those designers who only consider this strategy in some designs are involved with fast fashion brands that produce very low-price-pointed products. In those designs, their prime objective is to reduce the cost, and the retailer does not expect a long product life. This evidence thoroughly pushes the industry to rethink the marketing concepts that emerged during the 2010–2020 period to increase the speed of the ecosystem of the fashion industry, mainly the fast fashion concept, which produces multiple collections within a short period.

4.6 Design for Optimising Functionality

Considering functional values is a crucial requirement of product designers; it is reflected in the responses, too. None of them have any idea about this strategy. However, 46% said they sometimes considered this strategy (Fig. 9). Therefore, it was questioned during the interviews. The main reason not to consider the optimisation

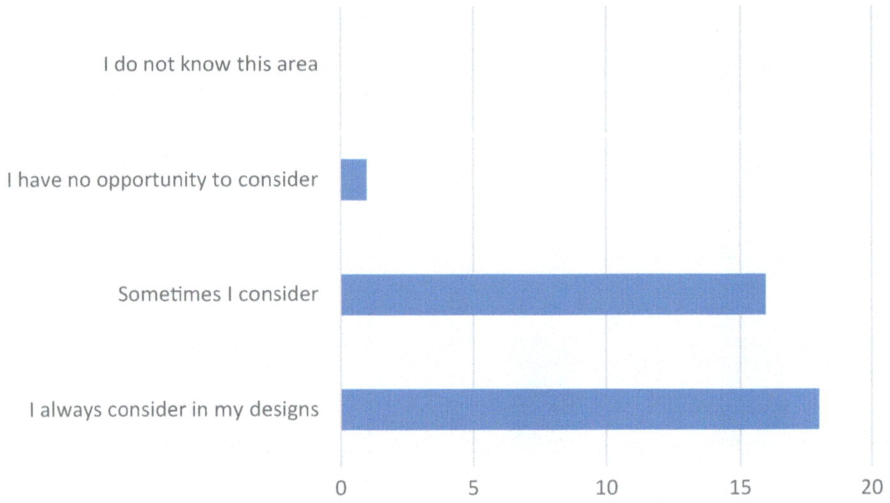

Fig. 9 Design for optimising functionality

of functional values in all the products is the nature of the mass market design approaches. It has several basic products in every range, and the designer pays less attention to optimising the functionality. Also, they pay more attention to the cost or appearance than the functionality in certain circumstances. "I am dealing with basic products, and the retailers always seek to reduce prices every season. Therefore, my major concern is not the functional values, but the cost…" Further, it is revealed that the designers lack awareness about considering functional values to improve sustainable factors.

4.7 Design for Reuse Recovery and Recycling

The insights related to this strategy could have been more questionable because at least 20% of the sample said they designed their products to be reused, recovered, or recycled (Fig. 10). However, mass-market products with these features can rarely be found. Therefore, this finding was thoroughly questioned during the interviews, and significant facts were revealed.

The main fact that was revealed was, the designers misinterpreted this strategy. Reusing material to produce a design or using recycled materials for a product comes under strategy 1: improving material efficiency. In this strategy, designers should pay different attention to designing a product to be reused, recovered, or recycled after the primary use. However, all the interviewees agreed that their previous understanding must be corrected. However, particular positives were revealed during the discussions. One interviewee with 18 years of experience in the intimates industry said they remove the brand identities and sell the proto samples to the employees of

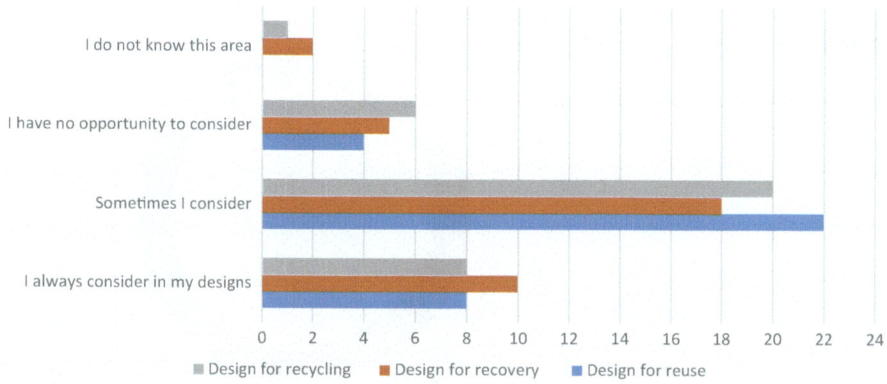

Fig. 10 Design for reuse recovery and recycling

the product development centre without sending them for landfilling. However, it is revealed that their approach needs to be communicated to the retailers. That approach is a good initiative from an environmental perspective. However, the ethical aspect should be improved by raising awareness among the relevant stakeholders because certain retailers have their brand protection policies.

4.8 Avoid the Hazardous Substances

The chemicals are heavily used in the export apparel industry for many applications. However, the results show that the majority (63%) of the sample does not fully consider this strategy in their daily routines (Fig. 11). If the retailer asks them explicitly to avoid hazardous substances, they keenly pay attention by requesting information from the suppliers. Otherwise, they need substantial requirements to think about it. As mentioned in the other strategies, the reduction of cost factor is one of the critical requirements in this market segment, and designers put most of the effort into considering the cost rather than chemicals used in the materials or other applications such as prints, washing, and dyeing.

On the other hand, they depend on the suppliers' information and their certifications. "*My company only deals with certified suppliers; therefore, I do not need to consider it much in the design process…*" A designer who has 15 years of experience in the industry explained. As he mentioned, most suppliers have certifications from relevant bodies about the usage of chemicals; however, when the cost factor comes to the mainstream, most suppliers have to deal with different chemicals with no certifications. Therefore, information transparency of the materials should be vital to be available to the designers when they are designing. On the other hand, this chemical-related information may not be understandable to the designers. Therefore, user-friendly methodologies are needed to make that understanding.

Fig. 11 Avoid the hazardous substances

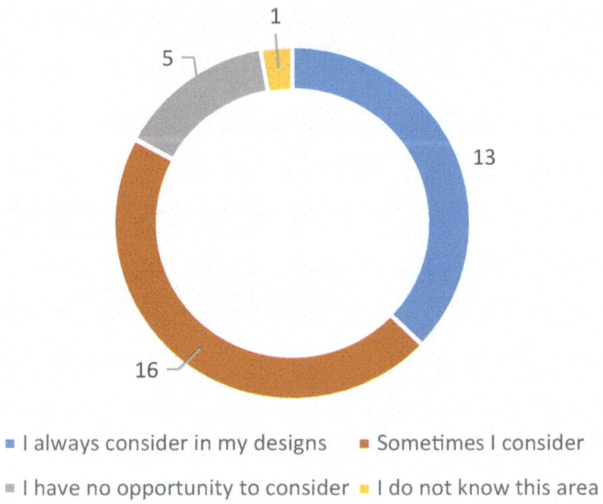

5 Discussion

The above findings reflect the imbalanced consideration of ISO 14062:2002 sustainable design strategies in the design process of the designers in the Sri Lankan apparel export industry. As visualised in Fig. 12, the primary consideration is given to the product's material efficiency, durability, and functionality. All the other strategies should be carefully considered, especially the economic use of land through product design, which gets the most minor consideration (57%) among all the other strategies. The main reason for poor consideration of land use is a lack of awareness about this design strategy.

Material efficiency, durability, and functionality of the product have a direct connection with the cost of the product. Material efficiency directly lies with material consumption for the product, which is calculated through marker efficiency and cutting performance. High marker efficiencies support price reductions. The durability factor is also essential to keep the brand value. Therefore, all the brands have durability-related quality specifications. Therefore, these industry-based designers must consider this factor to keep the retailers in their business.

On the other hand, as a designer, considering the functional values of the product offers high attraction to the buyers. Therefore, these three strategies are heavily used to sustain the business, ultimately supporting the environment. However, the purpose of using these design strategies is not centred on the sustainability angle but mainly on the profit angle, which indirectly supports sustainability.

However, the other design strategies should be strongly connected with the cost factor: then the designers would have more interest on using these strategies in their design routings. Converting non-renewable to renewable energy sources such as solar power is a comparatively high-cost investment. Therefore, it is easy to depend on conventional energy sources to maintain high-profit margins. Therefore smart ways

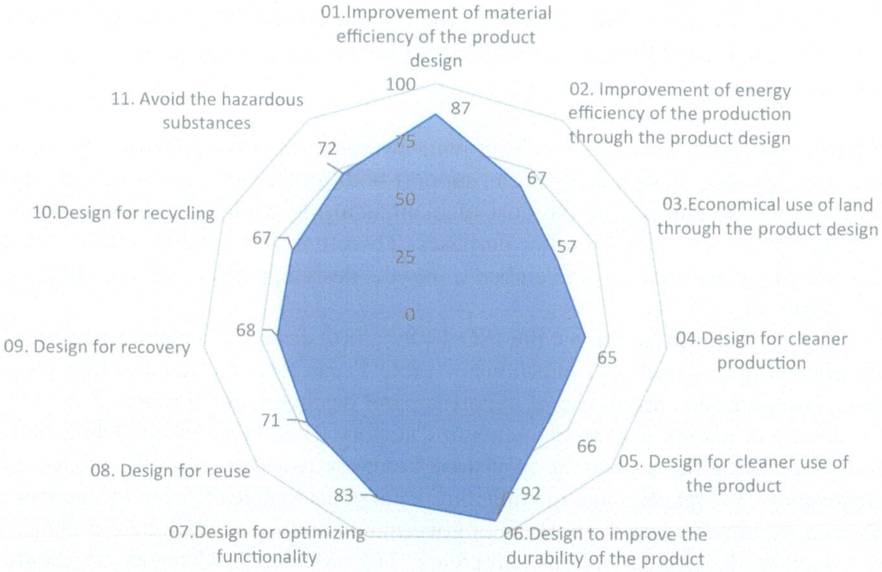

Fig. 12 Priority level of the designer for each design strategy

should be explored to the designer to introduce effective energy use as a design feature. Land usage is rarely discussed in the design community due to a need for more awareness about its importance in the design stage. On the other hand, designers need more resources and information to consider it in the design stage. Cleaner production also does not directly support the business's profitability; therefore, putting extra effort into making the cleaner production does not motivate the designers.

Furthermore, avoiding hazardous substances in the production process also plays the same role; hence, it has no solid connection to making profits. However, due to the policies of the retailers and the governments, many manufacturers consider investments in non-renewable energy sources, cleaner production methodologies, and avoiding hazardous substances. However, it is noticed that the companies conduct all these positive approaches to keep the buyers expanding the business rather than purely thinking about the sustainability of the environment.

Cleaner use of the product, design for reuse, recovery, and recycling do not have a direct connection with the cost and profit of the product; hence, they directly connect with the post-manufacturing stages of the product. At the moment, considering them in the design stage, offer only a sustainability-related benefit. However they can introduce new marketing plans to circulate the product lifecycle by collecting use products of their own brands. These kind of strategies are still not widely use. It be the reason that, most of the designers need more awareness and concentration towards those strategies. However, at the moment, due to the strict rules and regulations of the governments, retailers need to show their commitment towards the sustainability of post-manufacturing stages. Therefore, designers need to show their competency in this area. However, the current competency in handling post-manufacturing related design strategies could be much higher, and they need extra support to improve it. Training programmes, user-friendly new tools, and practical exposure will support this requirement.

Furthermore, it is identified that improving awareness about considering the use of land in the product design process can support both profitability and sustainability. Less use of land can reduce the cost of manufacturing while allowing the world to utilise the excess land for other purposes. Therefore, it is vital to consider this design strategy and find ways to embed it into the design process for more effective consideration.

Based on these observations, the ISO 14062:2002 design strategies-based areas investigated in this study (11 questions in Table 1) can be classified into four categories based on the current use of export apparel designers in Sri Lanka (Fig. 13). The designers mainly use the elements in Category 01 to make profits while indirectly supporting the sustainability vision. Category two does not offer high-cost effectiveness, but retailers and manufacturers put considerable effort to escape from the negative impacts that the policies of government and non-government organisations such as the United Nations can create. The elements in Category 03 should be more strongly considered in the current context. However, the government's upcoming policies and regulations have started pressuring retailers and manufacturers to consider them in the design process. However, the support of the current facilities to consider these factors in the design process should be much higher than now. Therefore, new approaches, tools, and methods may be needed to fulfil this requirement. Category 04 needs to be considered by improving the awareness; hence, considering the effective use of lands during the design process can support both profitability and sustainability.

6 Conclusion

The textile and apparel industry is the second largest polluter in the world and the largest export industry in Sri Lanka; hence, it is vital to investigate its impact on SDG12: ensure sustainable consumption and production patterns to find possible regulatory frameworks. Also, the literature proves that the design stage is the critical phase to be regulated to balance sustainability in apparel production and the consumption stages. ISO 14062: 2002 offers eight sustainable design strategies to be used in the product design process. Therefore, this study was conducted to see the current use of these strategies in apparel designers in Sri Lanka. The study was conducted through a mixed approach; qualitative information was gathered and converted to quantitative values for a comprehensive analysis. Ten well-experienced designers from the sample were further interviewed to clarify the patterns created through the quantitative analysis.

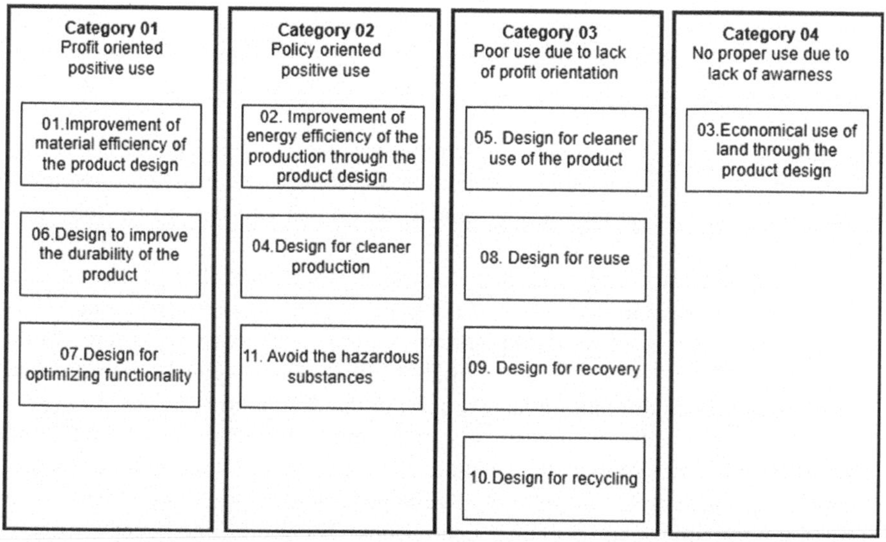

Fig. 13 Classification of ISO 14062 design strategies-based areas investigated in this study

The study revealed that, the export apparel designers in Sri Lanka use ISO 14062:2002 sustainable design strategies in two ways: profit-oriented and policy-oriented perspectives. Also, specific strategies still need to be considered thoroughly due to a lack of profit orientation in those strategies. However, there is a need to implement those strategies due to the pressure from government-level policies. Also, the lack of awareness of the effective use of land needs to be addressed. However, it can make an impact on both profit and policy orientations. Based on the findings, the following recommendations are made.

1. Educating the designers on generating profits and existing policy-oriented design approaches in Category 02 is necessary. Therefore, retailers and manufacturers should offer extra freedom to the designers to use those strategies in the design process more conveniently.
2. The areas in Category 3 should be thoroughly considered with government policies to pressure the retailers to use them effectively. At the same time, there should be convenient tools and methods for designers to implement those strategies in the design process. More resources are needed to fulfil that need. Furthermore, training and awareness sessions can be used to educate the designers.
3. It is essential to conduct further studies to see how those areas in Category 03 can make profits; then, the retailers will offer more support to implement Category 03 effectively than current practice.
4. It is essential to revisit the ISO 14062:2002 design strategies and refine strategy 04: a design for cleaner production and use because production belongs to the manufacturing phase, and use belongs to the post-manufacturing stage. When we take both areas together, it is not very easy to evaluate the performance because

the attention level of the designer in the manufacturing stage (profit-oriented) and post-manufacturing stage (policy-oriented) is different. Dividing strategy 04 into two sections may support to find tailor-made solutions that offer robust outcomes.

5. Similarly, strategy 07: design for reuse recovery and recycling should be revisited since all three categories have identical and complex scenarios. Discussing all three areas under one basket may hinder the crucial situations; therefore, it is better to divide strategy seven into three sections.
6. It is identified that strategy 03, the economical use of land through the product design, is neglected due to the lack of awareness regarding the true meaning of this strategy; it is recommended to implement comprehensive awareness programmes for designers that highlight the potential of this strategy to gain both profits and policy-level gains through implementation to the design process.
7. It is not only awareness but also very important to introduce user-friendly and practical tools that support the designer in balancing profit and policy orientations. Designers are more into creative approaches and reluctant to deal with numeric and statistical analytical approaches. Therefore, intelligent tools can support the designer in making such comfort.

Implementing these improvements can support achieving SDG12: ensure sustainable consumption and production patterns in Sri Lankan and the global apparel industry.

References

1. United Nations (1972) Report of the United Nations conference on the human environment—A/CONF.48/14/Rev.1
2. United Nations (1987) Report of the world commission on environment and development—our common future. United Nations. https://www.sswm.info/sites/default/files/reference_attachments/UNWCED1987BrundtlandReport.pdf
3. United Nations (2015) Transforming our world: the 2030 agenda for sustainable development. United Nations. https://sustainabledevelopment.un.org/content/documents/21252030AgendaforSustainableDevelopmentweb.pdf. https://doi.org/10.1007/s13398-014-0173-7.2
4. United Nations (2023) GOAL 12: sustainable consumption and production. https://www.unep.org/explore-topics/sustainable-development-goals/why-do-sustainable-development-goals-matter/goal-12#:~:text=Sustainable%20Development%20Goal%2012%20encourages,are%20toxic%20to%20the%20environment
5. Munasinghe PD, Druckman A, Dissanayake DGK (2021) A systematic review of the life cycle inventory of clothing. J Clean Prod 320
6. World Bank (2023a) Population density (people per sq. km of land area)—Sri Lanka. https://data.worldbank.org/indicator/EN.POP.DNST?locations=LK
7. World Bank (2023b) CO_2 emissions from liquid fuel consumption (kt)—Sri Lanka. https://data.worldbank.org/indicator/EN.ATM.CO2E.LF.KT?locations=LK
8. EDB Sri Lanka (2023) Export performance of apparel. https://www.srilankabusiness.com/apparel/about/export-performance.html
9. Ruwanpura KN (2022) Garments without guilt? Global labour justice and ethical codes in Sri Lankan apparels. Cambridge University Press. https://doi.org/10.1017/9781108937573

10. Jayasinghe J (2021) Energy and water consumption high in apparel sector. Print Edition. The Sunday Times, Sri Lanka. Sunday Times. https://www.sundaytimes.lk/210704/business-times/energy-and-water-consumption-high-in-apparel-sector-447767.html
11. Munasinghe PD, Dissanayake D, Druckman A (2021) An investigation of the mass-market fashion design process. Res J Text Apparel: 1560–6074. https://doi.org/10.1108/RJTA-08-2020-0089
12. Kerr J, Landry J (2017) Pulse of the fashion. Global Fashion Agenda & The Boston Consulting Group. http://globalfashionagenda.com/wp-content/uploads/2017/05/Pulse-of-the-Fashion-Industry_2017.pdf
13. Gwilt A (2011) Shaping sustainable fashion—changing the way we make and use clothes. In: Shaping sustainable fashion: changing the way we make and use clothes. Earthsacan. https://doi.org/10.4324/9780203126172
14. van Boeijen A, Daalhuizen J (2010) Delft design guide: design strategies and methods. Faculteit Industrieel Ontwerpen
15. British Standards Institute (2002) PD ISO/TR 14062:2002: environmental management. Integrating environmental aspects into product design and development. British Standards Institute
16. Black S (2008) Eco-chic the fashion paradox. Black Dog
17. Design for sustainability: a step-by-step approach. United Nations Environment Programme
18. United Nations Enivronment Management Group (2022) Delivering on the vision of the 1972 stockholm declaration and achieving the 2030 agenda for sustainable development A UN System contribution to Stockholm + 50 summary for policymakers. https://www.stockholm50.global/resources/documents
19. UNESCO World Water Assessment Programme (2021) The United Nations world water development report 2021: valuing water; facts and figures. https://unesdoc.unesco.org/ark:/48223/pf0000375751
20. Kostoska O, Kocarev L (2019) A novel ICT framework for sustainable development goals. Sustainability 11:1961
21. Reuters (2023) Despite climate pledges, fashion brands 'way off track' on cutting carbon from the catwalk. Reuters. https://www.reuters.com/sustainability/climate-energy/despite-climate-pledges-fashion-brands-way-off-track-cutting-carbon-catwalk-2023-07-31/
22. Gough I (2017) Heat, greed and human need: climate change, capitalism and sustainable wellbeing. Edward Elgar Publishing Limited
23. Breward C (2003) Fashion. Oxford University Press
24. Laver J (2002) Costume and fashion a concise history. Thames & Hudson
25. Niinimaki K et al (2020) The environmental price of fast fashion. Nat Rev Earth Environ 1:189–200
26. Cristofol Rodriguez C, Mcquillan K, Segarra-Saaavedra J (2021) Sustainability communication as a sales and social change tool: fast fashion vs slow fashion. REVISTA DE COMUNICACION DE LA SEECI. https://doi.org/10.15198/seeci.2021.54.e667
27. Muthukumarana TT, Karunathilake HP, Punchihewa HKG, Manthilake MMID, Hewage KN (2018) Life cycle environmental impacts of the apparel industry in Sri Lanka: analysis of the energy sources. J Clean Prod 172:1346–1357
28. Mishra AK, Rajput P, Singh A, Singh CK, Mall RK (2021) Effect of lockdown amid COVID-19 on ambient air quality in 16 Indian cities. Front Sustain Cities 3
29. Senarathna M et al (2021) Impact of COVID-19 lockdown on air quality of Sri Lankan cities. Int J Environ Pollut Remediation: 12–21. https://doi.org/10.11159/ijepr.2021.002
30. Sakamoto M, Ahmed T, Begum S, Huq H (2019) Water pollution and the textile industry in Bangladesh: flawed corporate practices or restrictive opportunities? Sustainability 11
31. Lglm E, Alwis AAPD (2021) Waste flow analysis in textile and apparel sector, a case study of textile and apparel wet processing industries in Sri Lanka. J Res Technol Eng 2
32. Kinkini Hemachandra DW (2015) Adoption of voluntary environmental practices: evidence from the textile and apparel industry in Sri Lanka. http://www.sandeeonline.org/uploads/documents/publication/1049_PUB_Working_Paper_93_Kinkini.pdf

33. Gwilt A (2012) Integrating sustainable strategies in the fashion design process: a conceptual model of the fashion designer in haute couture
34. Statista (2023) Apparel—Sri Lanka. https://www.statista.com/outlook/dmo/ecommerce/fashion/apparel/sri-lanka
35. EDB (2023) Sri Lankan apparel. https://www.srilankabusiness.com/apparel/
36. MAS (2023) Enriching the fabric of life. https://www.masholdings.com/
37. Wilson J (2001) Handbook of textile design: principles, processes and practice. Woodhead Publishing Ltd.
38. Sanches RA et al (2015) Organic cotton, lyocell and SPF: a comparative study. Int J Cloth Sci Technol 27:692–704
39. Hu J, Lu J, Zhu Y (2008) New developments in elastic fibers. Polym Rev 48:275–301. Preprint at https://doi.org/10.1080/15583720802020186
40. Yip E, Cacioli P (2002) The manufacture of gloves from natural rubber latex. J Allergy Clin Immunol 110:S3–S14
41. Çay A (2018) Energy consumption and energy saving potential in clothing industry. Energy 159:74–85
42. Yasin S, Behary N, Rovero G, Kumar V (2016) Statistical analysis of use-phase energy consumption of textile products. Int J Life Cycle Assess 21:1776–1788
43. Common Objective (2021) Measuring fashion's ecological footprint. https://www.commonobjective.co/article/measuring-fashion-s-ecological-footprint#:~:text=Currently%2C%20about%205%25%20of%20the,materials%20for%20the%20fashion%20industry
44. Sandin G, Peters GM, Svanström M (2013) Moving down the cause-effect chain of water and land use impacts: an LCA case study of textile fibres. Resour Conserv Recycl 73:104–113
45. Daniela K, Casallas A, Melisa Y, Cortés A (2018) Cleaner production as an innovative strategy applied in a textile company. 9
46. Nielsen KS, van der Linden S, Stern PC (2020) How behavioral interventions can reduce the climate impact of energy use. Joule 4:1613–1616
47. WRAP (2017) Sustainable clothing A practical guide to enhancing clothing durability and quality. http://www.wrap.org.uk/sites/files/wrap/SustainableClothingGuide2017.pdf
48. WRAP (2015) Sustainable clothing a practical guide to enhancing clothing durability and performance introduction: sustainable clothing guide. www.wrap.org.uk/textiles
49. Weerasinghe DU, Perera S, Dissanayake DGK (2019) Application of biomimicry for sustainable functionalization of textiles: review of current status and prospectus. Text Res J 89:4282–4294
50. Watkins SM (1981) Using the design process to teach functional apparel design. Cloth Text Res J 7:10–14
51. Osterley R, Williams I (2019) The social, environmental and economic benefits of reuse by charity shops. Detritus 7:29–35
52. Sandin G, Peters GM (2018) Environmental impact of textile reuse and recycling—a review. J Clean Prod 184:353–365
53. Bukhari MA, Carrasco-Gallego R, Ponce-Cueto E (2018) Developing a national programme for textiles and clothing recovery. Waste Manage Res 36:321–331
54. Masoudipour E, Amirian H, Sahraeian R (2017) A novel closed-loop supply chain based on the quality of returned products. J Clean Prod 151:344–355
55. Chapagain AK, Hoekstra AY, Savenije HHG, Gautam R (2005) The water footprint of cotton consumption value of water. http://waterfootprint.org/media/downloads/Report18.pdf
56. Patterson P (2020) Chemical circularity in fashion. https://www.laudesfoundation.org/en/resources/2020/05/chemical-circularity-digital.pdf
57. Kiron MI (2022) List of restricted substances for textile and clothing industry. https://textilelearner.net/list-of-restricted-substances-for-clothing/#:~:text=It%20is%20calculated%20that%20over,AZO%20dyes%2C%20chromium%20and%20formaldehydes

Sustainability and Textile Testing

S. N. Niles

Abstract Sustainability is the ability to carry out required tasks and meet required needs without compromising the long-term availability of the required resources and the long-term condition of the environment. Sustainability recognizes the finite nature of available resources and requires that steps should be taken to ensure that products and processes do not adversely impact the environment, society and the economy. The textile industry is one industry that has been known to have an impact on the environment and as a result on society. This has been seen in different sectors of the industry, particularly the colouration and yarn manufacturing sectors. Chemicals, which are widely used for the colouration and finishing processes, can be harmful to the environment. The dwindling availability of raw materials warns against the indiscriminate use of textile materials. The substitution of synthetic materials in place of natural raw materials can cause problems about degradability. Testing plays a major role in ascertaining the suitability of materials and processes in terms of their sustainability and their impact on the environment. This chapter discusses various tests that can be carried out to ascertain the sustainability of textile materials and related processes. It also seeks to discuss the actual testing itself in terms of its own sustainability.

Keywords Sustainability · Textile testing · Quality · Performance standards · Environmental impact

1 Introduction

The textile industry plays a significant role in the global economy. Textiles play a major role in providing apparel for wearing, household textiles and a variety of material for a plethora of other applications. In Sri Lanka the textile and apparel sector accounts for the largest export industry and the highest foreign exchange earning industry in the country. Population growth around the world, greater awareness of

S. N. Niles (✉)
Department of Textile and Apparel Engineering, University of Moratuwa, Moratuwa, Sri Lanka
e-mail: niles@uom.lk

G. M. Ranathunga (ed.), *Textiles of Sri Lanka*, SDGs and Textiles,
https://doi.org/10.1007/978-981-97-0569-6_8

fashion trends and upward mobility vis-à-vis the standard of living have resulted in the greater overall consumption of textiles worldwide. However, over the years the textile industry has faced increasing scrutiny due to the adverse impact it has on the environment. High consumption of energy, water and chemicals is major causes for concern. The textile sector is responsible for large-scale air, water and solid waste pollution of the environment. Resource depletion, pollution and greenhouse gas emission have led to the textile industry having a negative impact on the environment [1]. It has been found that greenhouse gas emissions exceed 3.3 billion metric tonnes per year [2].

The dyeing and finishing sector is a major culprit with regard to water pollution, with wastewater from the dyeing process being a major contributor. Many of these dyes are toxic and can have detrimental effects on both plant and animal lives. In addition, many synthetic dyes are found to have carcinogenic and allergenic properties and constitute a threat to the health not only of the user but also of those working in the dye plants. These include azo dyes which have in the past been used to dye cotton and even disperse dyes, used to dye many synthetic fibre-based fabrics [3].

In yarn manufacturing, too, loose fibres can result in particles that may cause respiratory problems when inhaled and land and water pollution when emitted to the environment. In addition to all this the large volume of energy and water consumption during the manufacture of textiles and textile-based products result in a depletion of those resources.

All of these have prompted the need for sustainability in textiles. Textile testing for sustainability has emerged as a necessity, ensuring that textiles meet stringent eco-friendly criteria while maintaining quality and performance standards. This chapter deals with how textile testing could be used to achieve sustainability goals 12, 13 and 14.

2 The Importance of Sustainability in Textiles

Sustainability in textiles encompasses several key aspects:

Resource Efficiency: Sustainable textiles aim to minimize resource consumption. Key resources include water, energy and raw materials. In order to achieve this goal efficient production processes and recycling initiatives are of paramount importance.

Reduced Chemical Use: A number of the chemicals used in dyeing and finishing processes are prone to have detrimental effects on both the environment and human health. Sustainable textiles prioritize the use of non-toxic, low-impact dyes and finishes.

Recycling and Circular Economy: A sustainable textile industry promotes recycling and upcycling (which is the creative reuse of by-products and waste materials into new materials or products which will have value to the user), thereby reducing the amount of textile waste sent to landfills.

Supply Chain Transparency: Sustainability involves transparency throughout the supply chain, from sourcing raw materials to manufacturing and distribution. This ensures that ethical and environmental standards are met at every stage.

Durability and Longevity: Sustainable textiles are designed to be durable and long-lasting, reducing the need for frequent replacements and minimizing waste.

Testing can be carried out to ascertain to what degree most of these aspects can be achieved. Testing can help to checking that the products emanating from processes which have been modified to ensure better resource efficiency still match up to user requirements. For example, a process that uses less water but does not provide the necessary colour fastness may not be acceptable to the user. Testing helps to ascertain the levels of fastness obtained. Similarly, the quality of recycled and upcycled products can be checked by testing. Whether the textiles which are claimed to be sustainable have adequate durability is also ascertained by carrying out the required testing processes.

3 Textile Testing for Sustainability

Testing textiles for sustainability is to evaluate textiles to ensure that they meet specific eco-friendly and performance criteria. Such testing plays a pivotal role in guaranteeing that sustainable textiles not only fulfil environmental objectives but also maintain quality and functionality.

Fibre and Material Testing: Fibres are the basic building block of textile materials. Similarly other materials used for textile-based products, such as yarn and fabric, will largely depend on the fibres of which they consist. Therefore it can be fairly said that sustainable textiles begin with the selection of eco-friendly fibres. The eco-friendliness of a fibre will depend on many factors—the amount of resources it uses for manufacture, the substances used in its processing, its biodegradability and the impact of its waste on the environment. Testing could help to ascertain some of these. For example, fibres which have poor abrasion resistance or strength are more liable to break and be removed from the material that they constitute, resulting in adverse impacts on the environment. Fibres which are not biodegradable will cause serious problems in terms of solid waste, while those which require toxic chemicals will also cause pollutions even if they degrade and may cause harm to the wearer as well as pollution to the waterways.

Dyeing and Finishing Processes: Sustainable dyeing and finishing methods are examined to determine their environmental impact. Textile testing assesses factors such as colour fastness, the use of non-toxic dyes and the presence of chemicals and other substances that could have undesirable effects on the human body and the environment in general. Effluents from the dyeing and finishing processes, if not properly treated but discharged into the water bodies, may cause the temperature and pH of the water to change to levels that are harmful to aquatic life [4].

Performance and Durability Testing: Sustainable textiles must meet performance standards to ensure they are fit for their intended purpose. In addition they should be such that they will have longevity in use and will not deteriorate in a manner that will cause harm to the environment. For example, a fabric which tends to low abrasion and pilling resistance, or which shed fibres easily during use, may not only cause pollution to the immediate environment, but also during cleaning may shed large volumes of fibre into the water, which will subsequently cause water pollution as this water is discharged into the sewers and beyond. Testing of tear resistance, abrasion and pilling resistance, snagging and other related factors serve to evaluate the durability of the textile.

Environmental Impact Assessment: Life cycle assessments and environmental footprint analysis help quantify the environmental impact of a textile product, from raw material extraction to disposal. Sustainable textile testing laboratories use life cycle assessments to identify areas for improvement and guide the manufacturer into eco-friendly design choices.

A number of testing protocols publish lists of restricted substances, of which the presence in excess of specified limits can result in potentially harmful residues. These include azo dyes, dimethyl formamide (DMF), dimethyl fumarate (DMFu), formaldehyde, cadmium, chromium, nickel, mercury, lead and phthalates, among others. The testing of emissions from chemicals used is also carried out. An example of such a list is that published by the American Apparel and Footwear Association [5], which lists a plethora of substances under different categories: disperse dyes, flame retardants, formaldehyde, heavy metals, gases, pesticides, phthalates and a variety of categories of other chemicals. A number of test methods are available to check these emissions, and a few of them are as follows:

- Formaldehyde release—ISO 14184, which helps determine the amount of formaldehyde. Part 1 of the standard specifies a method to find out the amount of free formaldehyde and formaldehyde extracted partly through hydrolysis by means of a water extraction method [6]. Part 2 refers to formaldehyde released under accelerated storage by way of vapour absorption [7]. ISO/FDIS 14184–3, which is under development, specifies how to find the same using liquid chromatography [8].
- pH—ISO 3071, specifies the determination of the pH of an aqueous extract of textiles [9]. pH can also be measured using AATCC TM81 [10].

Textile materials that are no longer usable are often thrown out and end in a landfill. However the biodegradable nature of the material is a matter of concern. Biodegradability tests can be carried out using soil to check the degradation as well as the possible adverse impact on the soil due to degradation. A number of organizations carry out such testing using test methods such as ISO 11721, AATCC 30 and derivatives of these and others.

Another aspect of environmental impact is the release of microfibers from synthetic materials into the environment. This can happen during fabric manufacture as well as during the cleaning of garments. Such fibres often bypass the effluent treatment process and can cause harm to waterways and to aquatic life. Test methods such as AATCC TM 212 [11] are used to quantify microfiber shedding due to accelerated laundering. This method claims to determine fibre shedding not only for synthetics but also for natural fibres and blends [12]. Another similar test method is DIN SPEC 4872. This method tests a number of parameters: fibre release during washing, biodegradability of the textile discharge and the effect the residue has on the water quality in terms of ecotoxicology [13].

Regulatory Compliance: Sustainable textiles must comply with various international standards and certifications, such as OEKO-TEX, Global Organic Textile Standard (GOTS), IVN Best, Worldwide Responsible Accredited Production (WRAP) and Organic Content Standard (OCS), to name a few. OEKO-TEX Made in Green is a product label that certifies that textile and leather products have been manufactured in environmentally friendly facilities under safe and socially responsible working conditions, of materials that have been tested for harmful substances. The tests for harmful substances are carried out under stringent criteria specified under the OEKO-TEX Standard 100 Criteria Catalogue [14] and according to testing procedures defined under the OEKO-TEX Standard 100 Testing Methods [15].

4 Sustainability in Textile Testing

Whereas testing textiles for sustainability is gaining more and more importance and whereas a number of organizations are making efforts to develop test methods and testing protocols and carry out the same in order to assess and certify the sustainability of textile materials and processes, there is another side to this that needs to be given serious consideration. This is the whole aspect of ensuring sustainability in the testing process itself.

Textile testing does not happen in a vacuum. The testing process itself consumes a large amount of energy, water and chemicals. In addition some of the chemicals used are of a hazardous nature. As such it is important to assess the sustainability of the textile testing process itself.

Testing and analysis of textiles is generally done within a conditioned atmosphere. This is due to the hygroscopic nature of textiles, which exchange moisture with the atmosphere and are thereby subject to variable properties when conditions differ. It is usually recommended that testing be carried out in standard testing conditions of 20 ± 2 °C of temperature and $65 \pm 4\%$ relative humidity [16]. For a number of tests, however, a pre-conditioning process is carried out prior to conditioning to ensure that the test material reaches equilibrium from the dry side. The recommended conditions for pre-conditioning are 10–25% relative humidity and up to 50 °C temperature. To maintain these conditions laboratory conditioning units are needed, which will require the use of large amounts of electricity and water. This is further exacerbated by the amount of electricity needed for the various equipment used. For instant the standard lamp in a Xenon Arc light fastness tester has a rating of around 2500 W.

The test for colour fastness to light may require the lamp to burn for 30 h at a time. Similarly a fume hood is said to consume around 3.5 times more energy than an average house [17].

Water consumption is another factor that needs to be taken into consideration. Several textile tests require a large volume of water consumption: dimensional stability to washing, colour fastness to washing, water repellency tests and appearance after washing, just to name a few. In addition to these a number of other tests require water-based solutions to be prepared or require some amount of water as part of their process. Such tests, when in high demand, will necessitate large-scale water consumption.

Water repellency testing has become increasingly sought after in recent times [18]. It has been claimed that a typical tester running for eight hours a day over a five-day week will consume 3360 L of water over that period [18]. However it is also claimed by certain manufacturers that by recycling this water they are able to save 83% of this consumption.

Another point of concern is the hazardous chemicals that may be used for testing. For example, for a long time dimethyl formamide (DMF) has been used to identify fibres using the solubility test [19]. Exposure of analysts to the hazardous chemicals they test is also a point of concern.

A lot has been said of waste disposal in textile manufacturing and textile finishing. However it is doubtful that the same care is taken for waste disposal of consumables, waste water and chemical discharge from testing is carried out. Often consumables such as rubber diaphragms from the bursting strength test, abradant and backing material from the abrasion resistance tests and tested specimens are often discarded into the normal waste, without a thought as to its ultimate destination and impact on the environment.

Yet another source of waste of material in fabric testing is the fact that, in order to ensure that test results are as representative of the sample tested, often specimens are spaced out across the fabric. This results in large portions of fabric being wasted.

5 Challenges and Future Directions

While sustainable textile testing has made significant strides, several challenges remain:

Standardization: The lack of standardized testing methods can make it difficult to compare and evaluate sustainable textile products effectively. The industry must work towards consistent testing protocols.

Cost: Sustainable textile testing can be expensive, especially for small and emerging brands. Lowering testing costs and increasing accessibility are crucial.

Innovation: Continuous innovation is essential to develop new test methods and processes which are sustainable in themselves, while maintaining quality and performance standards. A simple example would be to try to combine tests using the same specimens, as is suggested in the AATCC test method for dimensional stability to laundering, where it is suggested that the same specimens be used to test for change in skewness.

In conclusion it can be said that sustainable textile testing is an integral part of the textile industry's journey towards sustainability. As technology and innovation continue to advance, the textile industry can look forward to an increasingly sustainable and responsible future.

References

1. Shiwanthi S, Lokupitiya E, Peiris S (2018) Evaluation of the environmental and economic performances of three selected textile factories in Biyagama export processing zone Sri Lanka. Environ Dev 27:70–82
2. Quantis (2018) Measuring fashion: insights from the environmental impact of the global apparel and footwear industries. Full report and methodological considerations. Available online https://quantis.com/wp-content/uploads/2019/11/measuringfashion_globalimpactstudy_quantis_2018.pdf
3. Pizzicato B, Pacifico S, Cayuela D, Mijas G, Riba-Moliner M (2023) Advancements in sustainable natural dyes for textile applications: a review. Molecules 28:5954. https://doi.org/10.3390/molecules28165954
4. Radhakrishnan S. Handbook of sustainable apparel production, 1st edn. In: Muthu SS (eds) Chapter 14. Eco-parameters and testing of sustainable textiles and apparels. CRC Press, Taylor & Francis group
5. AAFA Restricted Substance List (RSL), Version 23, American Apparel & Footwear Association, Feb 2023
6. ISO 14184-1:2011. Textiles—determination of formaldehyde—part 1: free and hydrolysed formaldehyde (water extraction method)
7. ISO 14184-2:2011. Textiles—determination of formaldehyde—part 2: released formaldehyde (vapour absorption method)
8. ISO/FDIS 14184-3. Textiles—determination of formaldehyde—part 3: free and hydrolysed formaldehyde (extraction method) by liquid chromatography
9. ISO 3071:2020. Textiles—determination of pH of aqueous extract
10. AATCC TM 81-2022. Test method for pH of the water-extract from wet processed textiles
11. AATCC TM 212-2021. Test method for fibre fragment release during home laundering
12. Wyman D (2021) AATCC method quantifies fibre shedding. AATCC Newsletter, Aug 2021
13. DIN SPEC 4872: 2023. Test method for textiles—determination of fibre release during washing and aerobic degradation level in aqueous medium in consideration of ecotoxicity
14. OEKO-TEX® STANDARD 100 Edition 02.2023. International Association for Research and Testing in the Field of Textile and Leather Ecology
15. Testing Methods STANDARD 100 by OEKO-TEX®. International Association for Research and Testing in the Field of Textile and Leather Ecology
16. ISO 139:2005/Amd 1:2011. Textiles—standard atmospheres for conditioning and testing
17. Mills E, Sartor D (2005) Energy use and savings potential for laboratory fume hoods. Energy 30(10):1859–1864. https://doi.org/10.1016/j.energy.2004.11.008
18. Montalto P (2022) The resurgence of water repellency testing. Textile World, Jan/Feb 2022. Textile Industries Media Group, Georgia
19. AATCC TM 20A: 2020. Test method for fiber analysis: quantitative

An Investigation of Fashion for Career Development in Women's Role in the Apparel Industry in Sri Lanka

M. A. H. S. Hansadhi, D. S. Wijerathne, and C. Thenuwara

Abstract The apparel industry in Sri Lanka has made a significant contribution to the nation's economic development and gives lots of job opportunities for women's empowerment. However, the apparel industry still has issues with gender discrimination and the reluctance to give chances to get the advancement of women into senior roles. The purpose of this study is to examine how fashion plays a supportive role in women's carrier development in the Sri Lankan apparel industry. This study will specifically look at the connection between wardrobe preferences, professional image, and prospects for career growth for women in the apparel industry. A survey questionnaire and semi-structured interviews will be utilized as part of a mixed-methods strategy to conduct the activity. The female employees in high-ranking positions in the Group of MAS and Brandix in the Sri Lankan apparel sector are investigated through the survey by setting up a questionnaire in order to gather information on their fashion preferences, opinions about their professional image, and career aspirations. A selection of survey respondents who have successfully advanced in their careers in the sector will participate in semi-structured interviews. Their techniques for success and experiences utilizing fashion as a tool for job growth will be covered in the interviews. This study makes the hypothesis, based on previous studies, that there is a positive connection between a woman's recognized professional image in the Sri Lankan apparel industry and her fashion choices, which in turn has a beneficial impact on her career growth chances. According to the study, women who intentionally use fashion to improve their professional image are more

M. A. H. S. Hansadhi
Faculty of Graduate Studies, University of the Visual and Performing Arts, Colombo, 07, Sri Lanka
e-mail: hshansadhi@gmail.com

D. S. Wijerathne (✉)
Department of Textile and Apparel Technology, The Open University of Sri Lanka, Nawala, Nugegoda, Sri Lanka
e-mail: dswij@ou.ac.lk

C. Thenuwara
Department of History and Theory of Arts, Faculty of Visual Arts, University of the Visual and Performing Arts, Colombo, 07, Sri Lanka
e-mail: chandraguptha.thenuwara@gmail.com

G. M. Ranathunga (ed.), *Textiles of Sri Lanka*, SDGs and Textiles,
https://doi.org/10.1007/978-981-97-0569-6_9

likely to be viewed as having project leadership personalities and are more likely to advance in their jobs than those who do not. And the study focuses on women's career development and empowerment in the Sri Lankan apparel industry, which is aligned with several Sustainable Development Goals such as gender equality, responsible consumption, and production. The results of this study will shed light on how fashion may support women's growth in Sri Lanka's garment sector. It is anticipated that the findings will help to shape policies and initiatives that promote women's career advancement and improve working conditions for women in the sector. The study will also offer suggestions and approaches that women can take to improve their professional image and possibilities for job progression. The research's unique selling feature is that it offers a thorough analysis of the relationship between fashion preferences, professional image, and career advancement for women in Sri Lanka's garment industry.

Keywords Feminine fashion · Career development · Female employees · Apparel industry · Sri Lanka

1 Introduction

The fashion industry inherently thrives on a perpetual demand for change. This dynamic quality, though, hinges on a specific prerequisite that must be met for it to become evident. Scholars have meticulously explored the intricate factors that facilitate the continuous transformation of fashion, putting forth various theoretical frameworks to elucidate these mechanisms. Simmel (1904) makes an important point about how cultural differences influence how people dress. According to [18], fashion serves two symbolic purposes by defining personal identities and encouraging a sense of community. Veblen [29] asserted that women, historically considered as avid shoppers, symbolize the economic prowess of their male counterparts through their clothing choices, further demonstrating the centrality of fashion to women's social standing.

Theoretical frameworks within the realm of fashion often incorporate a psychological dimension, with a significant emphasis on fundamental human instincts such as the strong desire for social validation. People's need to promote and enhance their self-esteem is frequently highlighted as a particularly strong motivating factor. The fashion industry has a significant capability for producing profound effects in the search of self-expression, effectively serving as a concrete medium for grabbing attention. In the past, clothing has sparked a great deal of interest and discussion in a variety of academic fields, including social science, psychology, anthropology, art, history, and cultural and communication studies. However, its impact permeates numerous aspects of human existence and goes far beyond the confines of these scholarly endeavors. Fashion is innately dynamic, evolving in lockstep with the relentless march of time. It continuously reflects the prevailing zeitgeist of society, embodying the modern ethos. A wide range of academic disciplines, including social science, psychology, anthropology, art, history, as well as cultural and communication studies, have been perpetually enthralled by attire, which is intricately related to the overarching tapestry of fashion. Nevertheless, it has a pervasive impact that permeates and protects the entirety of human existence. Fashion is known for its incessant change, which keeps it in a constant state of flux while reflecting the changing sensibilities of the time.

- The term "Fashion" encompasses a diverse range of disciplines, and notable perspectives are emphasized in relation to the areas of study: "The purpose of fashion can be as simple as you need clothing to maneuver in the world. Forget fashion as a trend, think of fashion and clothing as a necessity. You need a pair of pants, a shirt, and shoes to go on that interview, to the grocery store, to your cousin's wedding. Its' how we present ourselves to live" [10].
- "Human language underlies the meaning and the word of fashion spreads it, establishing its reality. When speaking of fashion, such a position remained valid until the late 1970s—the verbal text dominated the visual. In contemporary society, fashion relies on visual communication" [23].
- "Our clothing is the physical representation of our perceptions, our dissatisfactions, and our desires. When we look beyond the physical to our internal workings, we can create a change at the core. Unlike change that occurs in therapy, these difficult internal examinations are softened by the light of the wardrobe makeover... Taking care of your self begins with self-discovery. The clothing you put on your back is an incredibly accurate indicator of what you think of yourself and your life. Cracking open the closet doors can lead to great insight. When you strive toward self-discovery, improvement often follows wearing clothing that makes you feel comfortable, happy, and good about yourself does make life better" [1].
- "Fashion is not a trivial and ephemeral phenomenon is the way in which fashionable clothing and have been used to express and shape personal and social identities" [2].

The preceding explanations clearly demonstrate that an individual's outward appearance in communal settings reflects their internal essence. The initial encounter with someone's physical presentation leaves a significant impact. The attire of workers, particularly females, could potentially influence their professional roles.

According to [31], vocations have historically been viewed as a purposeful progression via a series of connected occupations. Women think that they should have the same oppotunities as men to climb the corporate ladder. They are following the same procedures and skills to build up their career development. When comparing the previous research, it reflects that there are no specific researchers based on career development and fashion. Through this research, it is giving a deeper understanding of women's career and fashion. According to [31], women believe that they have similar chances to climb the career ladder particularly if women pick careers like the men's jobs that have same capacity, goals, and aptitudes [31]. According to the previous career development studies, it has not directly addressed the research question, which is affected to female career development. When considering the viewpoint of the employee, career development career advancement has a prominent effect because it works as a useful method or even a road map for them to achieve the career goals which may include flexible schedule and job satisfaction with high salaries and other advantages like incentives and bonuses [31]. Here are several essential definitions of "career growth":

- "Career development is the process of implementing and developing the self-concept" [9].
- "Career development refers to a change or a series of changes that occur in an individual's career" [4].
- "If career is recognized as a life process, it is vital to identify and understand the human participation in this process. Life career phenomena do not exist without human involvement. People are their own actors in performing career developmental tasks, and they are the self-agents in shaping their lives" [5].

Sri Lanka's garment industry is a significant employer, particularly for women, and a substantial economic force in the nation [14]. In this study, the Sri Lankan clothing business is examined, emphasizing the crucial part that women play in the ready-to-wear market. Given the industry's significant contribution to Sri Lanka's foreign exchange profits and the fact that women make up the second-largest portion of this income stream, the study's conclusions are especially pertinent.

An integrated assessment of the literature, in-depth interviews with significant stakeholders, and secondary data analysis are all part of the study methodology. The results provided light on the chances and obstacles for female garment industry workers in Sri Lanka to progress their careers. The study also investigates the link between female career advancement and fashion trends.

The technique of the study is built upon the literature review, which also aids in achieving its main goal. The study aims to comprehend the complex relationship between fashion trends and female employees' professional growth by critically reviewing pertinent literature.

This study also underscores the empowerment of women within the Sri Lankan clothing sector and their progression in their professional careers. This aligns with several Sustainable Development Goals, including those related to gender equality and responsible consumption and production.

2 Background

Fashion serves as a paramount mode of human expression, encompassing multi-faceted facets of both internal and external communications. Recent research underscores the profound influence of clothing choices on individuals' behaviors and actions. It is evident, through earlier scholarly definitions, that fashion serves as a mirror reflecting an individual's lifestyle. In initial encounters, an individual's appearance etches a lasting impression, with particular significance for female professionals, where their attire can wield considerable influence on their career trajectories.

The significance of dress in the constantly changing business world extends beyond aesthetics and includes a sophisticated web of nonverbal indicators. The way you dress sends signals about your level of skill, power, and compliance with company standards. They can also represent a person's ingenuity, adaptability, and attention to detail. Female professionals' dress serves as a blank canvas on which they paint their professional personas. It may represent their dedication to their jobs, their desire for career growth, and their aptitude for navigating the nuances of corporate culture. Therefore, it is crucial to understand that fashion is more than just a matter of taste. For female professionals, it is a strategic instrument that they use deftly to negotiate the complex terrain of their working lives.

This relationship between style and professional success also extends to personal empowerment. Many female workers find that their clothes give them confidence and enable them to convey competence and authority. They can declare their identities and defy traditional gender roles in the workplace by using it as a form of self-expression. It is important to recognize that clothing is more than just a surface-level part of work life as we go deeper into the study of fashion and its impact on female career paths. It is a dynamic and powerful force that affects how female professionals see themselves and how others see them, which ultimately affects their career objectives.

It is impossible to emphasize the importance of fashion as a nonverbal communication tool for female workers in a time of increased awareness of diversity, inclusivity, and gender dynamics in the workplace. Beyond just the outward appearance, clothing captures a complex interplay between identity, empowerment, and social expectations. The clothing decisions made by female business executives strike a careful balance between self-expression, societal conventions, and corporate compliance. As a result, exploring the many facets of fashion in women's career paths not only deepens our comprehension of this complicated phenomenon but also provides insightful information about gender relations in workplaces. As a result, fashion becomes both a topic ripe for more scholarly investigation and a canvas onto which the changing narratives of female professionals in the modern industry are painted.

It is widely acknowledged that self-worth and self-assurance are pivotal to career advancement. External hindrances such as societal expectations and entrenched stereotypes undoubtedly impact an individual's success in the professional realm. Equally consequential are internal impediments such as self-esteem, confidence, and resilience [6]. An individual's sense of self is intricately molded by their experiences within the realms of work, relationships, and societal interactions [21]. In examining the determinants of female employees' career progression, current research endeavors to emphasize the significance of fashion as a catalyst for professional growth.

Recognizing the complex interplay between self-worth, self-assurance, and job advancement is essential given how these factors interact with one another. According to [6], a person's capacity to successfully negotiate the professional environment is largely dependent on their level of self-esteem. People with a solid sense of self-worth are better able to overcome obstacles and biases from the outside world, which helps them advance in their chosen fields of work. Additionally, a person's desire to take on new tasks, step up into leadership positions, and take advantage of possibilities for growth is greatly influenced by their level of confidence. The development of resilience is also essential because it gives people the tools they need to recover from setbacks and persevere in the face of difficulty. These internal elements, coupled with external cultural dynamics, influence how a woman's career develops through time.

The study's emphasis on women's professional advancement and empowerment within Sri Lanka's garment sector is in line with other SDGs, particularly those pertaining to gender equality and responsible consumption and production. This alignment of goals emphasizes how crucial it is to comprehend how a person's sense of self and decisions in clothing interact in the context of career advancement. According to [21], who said it beautifully, a person's sense of self is a rich tapestry woven from experiences across multiple domains, such as job, relationships, and societal interactions. These events affect a person's self-perception as well as how they show themselves to the outside world.

It is clear how crucial fashion is in a comprehensive perspective. Since clothing choices reflect and affect self-concept, they have a substantial impact on professional development. This dynamic interaction between self-identity and fashion preferences emphasizes the necessity to look at how fashion influences the development and empowerment of female employees in Sri Lanka's textile industry. In order to advance gender equality and sustainable development objectives, it is important to understand how women in the sector navigate their way to success. This convergence of self-concept, fashion, and professional development offers a unique opportunity to do so.

The Sri Lankan textile sector is crucial to the country's economy and offers a wide range of job possibilities. This study focuses on the ready-to-wear industry, which employs a sizable percentage of women employees and is Sri Lanka's second-largest source of foreign money. The ready-made clothes' business has a female workforce that makes up around 85% of its workforce, according to the 2017 Annual Survey of Industries [20].

3 Analyzing Workwear Fashion Ideals and the Concept of Professional Attire in Sri Lanka's Corporate Environment

The subtle dynamics of Sri Lanka's clothing standards and workplace fashion practices, as well as their possible effects on job performance, have received little consideration in historical research. The formality of clothes, its motivational impact, degrees of comfort, the role of conformity, and its overall impacts on productivity have all been carefully examined in the literature that has already been written. In an effort to increase output and efficiency, a number of countries and organizations have modified their dress rules at various points in time [11]. Clothing and appearance are powerful nonverbal communication tools that can convey a person's values, attitudes, interests, lifestyle, social connections, authority, position, and vocational identification in addition to serving as a means of personal expression (Sampson, Teachers' Perceptions of the Effect of Their Attire on Middle-School Students' Behavior and

Learning, 2016). Even though formal apparel is frequently linked to professionalism, identity, and prestige, it can occasionally unintentionally impede interpersonal relationships [28]. Contrarily, adopting more casual dress requirements as opposed to formal wear is seen as a way to reduce barriers to communication between managers and staff members and promote a feeling of community at work [28]. This viewpoint draws attention to the covert impact of dress on professional success.

Women in modern Sri Lanka have freed themselves from the confines of conventional domestic responsibilities and pursued rewarding occupations outside the home. In the past, women typically filled supporting positions because males were supposed to be the main providers of income. But the reality of today demonstrates a considerable transition, with women actively participating in a vast array of occupations. The combination of societal shifts and rising female workforce participation is blamed for this development. The beauty sector is booming as a result of women's changing societal role as full-fledged professionals.

The changing role of women in society is reflected in Sri Lanka's workplace dress scene, in conclusion. It exemplifies how industrialization and cultural changes have combined to acknowledge women as necessary employees. The evolution of office fashion has been influenced by the commercialization of femininity, allowing women to express their professional identities with assurance and self-assurance.

The way we present ourselves to others and how we view ourselves have a big impact on what we wear. Although wearing a suit and tie is frequently seen as a sign of respect and professionalism, it can also put up a psychological barrier between people. Conversely, casual attire can foster familiarity and warmth, but in some official circumstances, it can also be perceived as a lack of competence. We can learn more about the complex world of human behavior and social dynamics in diverse circumstances by looking at the psychological consequences of formal and informal attire [22].

Dellinger's further investigation in 2002 revealed that the clothes that people wore had an impact on how well they perceived themselves to be performing in their respective fields as editors and accountants. In magazine editing, creative staff members expressed feeling constrained by their professional appearance, finding it challenging to completely express their creative potential when dressed in a particular way. On the other hand, traditional staff accountants believed that dressing professionally at work increased their professionalism [7]. These results shed light on the nuanced relationship between female employees' preferences in clothing and professional progress.

3.1 Exploring the Evolution of Professional Attire Trends in Sri Lanka

In terms of design and appearance, workwear has changed greatly over time. Men and women used to be required to arrive at the office dressed in formal attire. But when the internet became more widely used in the 1990s, office attire began to become more relaxed and informal [30]. Business casual outfit, often known as conventional dress attire, formal wear, or formal attire, differs significantly from traditional business formal attire today.

The implementation of free-market economic concepts and globalization have made it simpler for Sri Lanka to adopt international design trends. This has helped the nation's apparel business, but it has also significantly increased the adoption of Western fashion styles, a tendency that started to gain momentum in the early twenty-first century [30].

In Sri Lanka's garment business, many administrative staff members who frequently deal with consumers prefer to dress in "smart casual" or semi-formal wear. However, a lot of employees have had issues with their uniforms. In Sri Lanka, there are not many options for appropriate professional attire, and there are also problems with cost, comfort, and a lack of fabric and fashion options [19].

These earlier findings emphasize how crucial it is to consider employees' dress choices because they have a big impact on customer interactions and can also have an impact on individual productivity. It is significant to remember that employees' attire frequently reflects their sense of fashion, which might affect how they make decisions.

Previous studies have stressed the significance of professional dress. For instance, employers that place a high value on their looks are perceived as being more responsible, trustworthy, and competent in addition to having the required education and experience [15].

Additionally, Solomon and [26] research indicates that fashion-conscious employees may be more diligent and productive. Office attire has changed from traditional business casual to a more relaxed and informal dress code since the 1990s, when the internet boom first took off [26].

It is crucial to remember that there appears to be a research gap in Sri Lanka regarding awareness and comprehension of the relevance of fashion in enterprises. Finding this gap and looking into how fashion plays a part in accomplishing organizational objectives in Sri Lanka could be advantageous for businesses and their staff.

In addition to the previously mentioned insights, this study project focuses on the promotion of women's careers and their empowerment within the Sri Lankan textile sector. It is vital to understand how many SDGs are compatible with this agenda, especially those pertaining to gender equality and ethical consumption and production. In order to promote sustainable practices and gender equality, two vital objectives for Sri Lanka and the global society, it is crucial to comprehend the role that clothes and fashion play in this context. Stakeholders and policymakers may benefit from understanding the relationship between fashion, female empowerment, and sustainable development in the Sri Lankan textile industry.

3.2 The Impact of Fashion on Professional Advancement

Employee perceptions and the resulting patterns of values and expectations have a significant impact on organizational culture [16]. According to a 2017 study by Katherine Karl titled "Employee Self-Perceptions on Workplace Attire in Turkey," what people wear to work might have an impact on how they perceive themselves. People frequently visualize the proper business dress to wear at the office as well as the corresponding behaviors [13].

Employee behavior at work may vary depending on what they are wearing. People may come across as more responsible, competent, and credible when they are dressed professionally, for instance. Kwon [15] study, which discovered that employees' feelings are influenced by their job dress, supports this [15].

Employees' shared values, precepts, and expectations make up an organization's culture. This notion is comparable to [16] description of organizational culture as the "Cohesive framework" produced by shared perspectives among personnel.

The mindset and feelings of employees toward their workplace make up the organizational culture. It is predicated on their common values, opinions, and goals. Depending on what they are wearing, people's views of themselves can change. This indicates that individuals often judge others based on their appearance and how they expect them to act. Employees who dress professionally at work are more likely to behave professionally. This is due to the fact that they are aware of the behavioral standards linked with their attire. Dressing professionally can give the impression that you are more responsible, competent, informed, professional, honest, reliable, intelligent, trustworthy, industrious, and efficient, according to Karl et al. [13].

According to Joy V. Peluchette's 2007 study, the impact of workplace wardrobe on employee self-perceptions, people have historically utilized their clothing as a tool to attain particular aims in their social interactions. For instance, formal workwear is frequently strategically employed to elevate one's position and win respect, as demonstrated by instances in which people have worn specific sartorial choices to accomplish their professional objectives. In order to investigate the connection between personal style and work success, this study assumes that women's career growth may be significantly influenced by their choice of clothing [12]. Women are expected to keep up with current fashion trends and display themselves in a fashionable way in a society where fashion is highly regarded and a key component of a woman's identity. While encouraging, this also raises the prospect that dressing properly can facilitate career advancement.

People frequently alter their appearance in today's society to affect how other people view them. This is particularly clear at work, as people dress appropriately for key occasions like job interviews. Employees can still show their personality while adhering to dress requirements at work by accessorizing with unique colors, patterns, and accessories. Additionally, some people decide to spend more money on clothing from well-known fashion labels as a means of expressing their social and economic standing [17].

Less research has been done on the influence of clothes on embodied activities, despite the fact that its effects on nonverbal communication and initial impressions have been extensively examined. People who engage in embodiment activities make use of their bodies to express and experience their identities. Making clothing choices that reflect one's cultural notions of femininity or masculinity is one technique to do this [8].

Additionally, research has found that those who are happy with their physical appearance tend to be less worried and more extroverted. This shows that what we wear may have an impact on how we perceive ourselves or how our self-concept is shaped [27].

Research on the intricate connection between appearance and self-identity is highly productive. We can learn more about the dynamics of clothing and self-presentation by examining the ways in which people utilize their clothing to express their identities and embody their ideals.

Customer satisfaction may benefit from employee uniforms. According to Nelson and [3], the attire that staff choose to wear can contribute to making a welcome and warm environment for consumers [3]. Expanding on this idea, Mishra and Mishra (2015) contend that, in addition to our own sense of self-worth, decisions about our physical appearance are frequently influenced by how others see us.

When uniforms and their effects on female employees are taken into account, the situation gets more complicated. Their capacity to express themselves through personal clothes may be limited by uniforms, which may have an adverse effect on how happy they are at work. On the other hand, this discussion backs up the compelling hypothesis that, within the organization chosen for this research, fashion can have a substantial effect in women's employees' career advancement. Fashion can boost self-satisfaction, general efficacy, and perhaps even serve as a springboard for career progress when it permits greater personal freedom in terms of wardrobe [3].

4 Methodology

In order to explore the role of fashion in connection to the professional growth of women working in the Sri Lankan clothing sector, this research project undertook both narrative and systematic reviews. The review mainly focused on academic works that were released in English-language publications between 2010 and 2022.

The literature search was launched using an internet database called Google Scholar. In the search process, the following keywords were used: "fashion," "career advancement," "female professionals," "garment industry," and "Sri Lanka." The studies that met the established inclusion criteria were as follows: studies that (a) look at how fashion influences job progression; (b) focus on female professionals; (c) are conducted in the setting of the garment industry; and (d) have gone through peer review before publication.

This study endeavors to examine the impact of fashion on the career advancement of women employed in the Sri Lankan clothing industry. To achieve this objective, the study will employ a narrative approach to explore how fashion influences the perspectives and professional aspirations of female workers within this sector. To fully comprehend the significance of fashion in this setting, a thorough review of scholarly literature will be done. The use of a narrative approach will allow for a full investigation of how fashion influences women's career pathways in the apparel business.

It is going to start by carefully going over the body of research on fashion as a type of nonverbal communication. This analysis will concentrate on how clothing conveys a person's identity, status, occupation, authority, values, attitudes, and interests. Sampson [24] foundational work will serve as the theoretical cornerstone for our investigation. We will also consider current studies on how apparel affects conduct, particularly in formal contexts [24].

The promotion of women's professions and their empowerment in the Sri Lankan clothing industry are the main topics of this study. It is consistent with a number of SDGs, such as gender equality and responsible consumption and production. By examining how fashion might affect women's roles and advancement in this field, we intend to contribute to bigger projects to attain these crucial sustainability goals.

This study will concentrate on Sri Lanka's significant garment sector, which employs a considerable proportion of women and is crucial to the nation's economy. This study's major objective is to investigate how women's perspectives, chances for career progression, and professional growth are impacted by fashion in this field. The only subject of this study will be women who work in the garment business in Sri Lanka.

Our study will take a multimodal approach to data gathering, including surveys, interviews, and focus groups, in order to fully grasp the significance of fashion in this context. We will be able to get qualitative information about the actual experiences of female workers in the Sri Lankan garment sector and the part that fashion plays in their career advancement by conducting interviews and focus groups at the same time.

The primary objectives of this study are also in line with a number of Sustainable Development Goals (SDGs), notably those that deal with gender equality and sustainable consumption and production. The study would advance the fulfillment of these SDGs by enhancing the professional growth and economic empowerment of women in Sri Lanka's garment sector.

To better understand how the fashion business affects the career trajectories of female employees in Sri Lanka's garment industry, I will utilize narrative analysis to examine data from multiple sources. By using narrative analysis, I will be able to spot recurring themes and patterns in the data, which will shed light on the experiences of women employed in this field.

The aim of this study is to enhance our comprehension of the role of attire in the workplace and its impact on the career progression of women in Sri Lanka's apparel industry. This research contributes to important Sustainable Development Goals, such as gender equality and responsible consumption and production, by focusing on the professional development and empowerment of women within this specific industry context.

5 Conclusion

The extensive data and compelling arguments presented in previous studies underscore the significant correlation between fashion and the career development of women. It can be argued that fashion, as the foremost nonverbal communication tool in human existence, wields substantial influence over career achievements. Consequently, this study not only addresses one of the most intriguing inquiries regarding the nexus between women's career success and fashion but also paves the way for further exploration in this field. In essence, this research contributes to the broader global objective of gender equality and environmentally sustainable business practices, while also serving as a springboard for additional investigations into the intricate interplay between fashion and the career paths of women.

References

1. Baumgartner J (2012) You are what you wear, 1st edn. Da Capo Lifelong Books, 3755, New York
2. Bovone L (2012) Identities through fashion: a multidisciplinary approach. In: Bovone AM (ed). Berg, New York
3. Bowen KN (2000) The effect of employee uniforms on employee satisfaction. Sage 41(2):86–95. https://doi.org/10.1177/001088040004100221
4. Brown D (2002) Career choice and development, 4th edn. Jossey Bass, USA
5. Collin A (2006) The future of career. J Vocational Educ Train 21. https://doi.org/10.1017/CBO9780511520853
6. Collins G (2009) When everything changed: the amazing journey of American women from 1960 to the present, 1st edn. Little, Brown and Company, New York
7. Dellinger K (2002) Wearing gender and sexuality "on your sleeve": dress norms and the importance of occupational and organizational culture at work. 20(2):3–25. https://doi.org/10.1007/s12147-002-0005-5
8. Dipboye SK (2008) Effects of charismatic content and delivery on follower task performance: the moderating role of task charisma conduciveness. Group Org Manag 33(1):77–106. https://doi.org/10.1177/1059601106291072
9. Huang Q (2006) The nature of women's career development: determinants and consequences of career patterns. Psykologiska institutionen, Department of Psychology. Stockholm University, Stockholm. https://doi.org/10.1037/e530032013-001
10. Jacques R (2018) What's the point of fashion? Repeller 15
11. Jayasooriya LB (2016) Clothing in Sri Lanka: cultural conformity and contradictory rejection of women's individuality. J Human Soc Sci 3(2):327–332. Retrieved from https://doi.org/10.2139/ssrn.3631744
12. Peluchette JV, Karl K (2007) The impact of workplace attire on employee self-perceptions. Human Resour Dev Q 18(3):345–360. https://doi.org/10.1002/hrdq.1208
13. Katherine Karl JV (2017) Employee self perceptions regarding workplace attire in Turkey. J Org Psychol 17(6):125. https://doi.org/10.33423/jop.v17i6.1517

14. Kelegama S (2009) Ready-made garment exports from Sri Lanka. J Contemp: 579–596
15. Kwon YH (1994) The influence of appropriateness of dress and gender on the self-perception of occupational attributes. Cloth Text Res J 12(3):33–39. https://doi.org/10.1177/0887302X9401200305
16. Lunenburg FC (2012) Power and leadership: an influence process. Int J Manag Bus Administr 15(1):9. Retrieved from https://api.semanticscholar.org/CorpusID:156109330
17. Maran T (2021) Clothes make the leader! How leaders can use attire to impact followers' perceptions of charisma and approval. J Bus Res 124:86–99. https://doi.org/10.1016/j.jbusres.2020.11.026
18. Michael Kaern BS (1990) Georg Simmel and contemporary sociology, 1 edn., vol 119. Springer, London
19. Niromi Seram GD (2022) Business attire: challenges for employees regularly in contact with customers in the Sri Lankan apparel industry. Res J Text Appar. https://doi.org/10.1108/RJTA-09-2021-0113
20. Organization IL (2020) Empowering female workers. Information on ILO Publications, Turin
21. Powell GN (2011) Women and men in management, 5 edn. Sage Publications, Los Angeles. Retrieved from https://doi.org/10.1111/j.1468-0432.2006.00298_1.x
22. Reilly A (2020) Crossing gender boundaries. In: Andrew Reilly BB (ed) Intellect, Bristol
23. Roalnd B (1990) The fashion system. University of California Press, California
24. Sampson EC (2016) Teachers' perceptions of the effect of their attire on middle-school students' behavior and learning. Walden University
25. Sampson EC (2016) Teachers' perceptions of the effect of their attire on middle-school students' behavior and learning. Walden University, ProQuest LLC, Minneapolis
26. Schopler MR (1982) Self-consciousness and clothing. Pers Soc Psychol Bull 8(3):508–514. https://doi.org/10.1177/0146167282083018
27. Tricia Widner Johnson SK (2007) Appearance management behavior and the five factor model of personality. Cloth Text Res J 25(3):230–243. https://doi.org/10.1177/0887302X07302982
28. Vandana Nath SD (2016) Dress codes and appearance at work: body supplements, body modification and aesthetic labour. ACAS Research Publications. Retrieved from http://www.acas.org.uk/index.aspx?articleid=4953
29. Veblen T (1998) The theory of the leisure class: an economic study of institutions (great minds series), 1st edn. Prometheus, New York
30. Wardhana DY (2020) Employees' attitudes towards jobs: does work attires matter?
31. White B (1995) The career development of successful women. Women Manage Rev 10:4–15

Eco-Friendly Dyeing Techniques for Banana Fibers from *Banana Application in the Sri Lankan Context.* Promoting Sustainable Coloration Methods

Subashini Balakrishnan, G. L. D. Wickramasinghe, and U. G. Samudrika Wijayapala

Abstract Sustainable natural dyeing is a process that aims to minimize the environmental impact of banana fiber dyeing while promoting the use of eco-friendly and renewable resources. This approach involves the use of plant-based dyes. This study explores the application of natural dyes for banana fiber, a versatile and abundant natural material. This study explores the potential of natural dyes derived from locally sourced materials such as banana pseudostem sap (*Musa acuminata*) extracted juice, jamun fruit (*Syzygium cumini*), coconut shells powder (*Cocos nucifera),* Ceylon green spinach (*Basella alba*), and Tree Turmeri (*Coscinium fenestratum)* dyeing for banana fibers. Each of these natural dye sources offers a wide range of pastel color palette with diverse colors in the dyeing process. Some of these dye sources, such as banana pseudostem sap juice and jamun fruit juice, possess unique chemical properties that can create interesting visual effects. One of the key aspects explored in this study is the mordant-free dyeing techniques for banana fibers. Banana fibers in mordant-free dyeing techniques have the essential ability to provide long-lasting colors, color fastness, and dye absorption without the need for mordants. By exploring the potential of natural dye sources and mordant-free dyeing techniques for banana fibers, this chapter aims to promote sustainable and vibrant coloration options in the textile and fiber industry in Sri Lanka. The extraction processes, dyeing techniques, and color fastness properties of these natural dyes are examined to assess their viability in textile applications. By providing comprehensive information on

S. Balakrishnan (✉)
Department of Textile and Apparel Technology, Faculty of Engineering Technology, The Open University of Sri Lanka, PO Box 21, Nawala, Nugegoda, Sri Lanka
e-mail: Subaedu88@gmail.com

G. L. D. Wickramasinghe
Department of Textile and Apparel Engineering, University of Moratuwa, Moratuwa, Sri Lanka
e-mail: dharmasri@uom.lk

U. G. Samudrika Wijayapala
Department of Textile and Apparel Engineering, University of Moratuwa, Moratuwa, Sri Lanka
e-mail: samu@uom.lk

G. M. Ranathunga (ed.), *Textiles of Sri Lanka*, SDGs and Textiles,
https://doi.org/10.1007/978-981-97-0569-6_10

selected natural dye sources of five (5) and their properties for dyeing banana fibers, this study aims to promote sustainable and environmentally conscious practices in the fiber and textile industry and economic opportunities for local communities in Sri Lanka. Understanding the potential of natural dyes offers opportunities for creating unique and eco-friendly textiles, enhancing the market appeal of banana fiber-based products, and contributing to a more sustainable future.

Keywords Banana fiber · Natural dye · *Syzygium cumini* · *Musa acuminata* · *Cocos nucifera* · *Basella alba* · *Coscinium fenestratum* · Textile industry · Color fastness · Color properties

Abbreviations

GDP Gross Domestic Product
K/S Absorption coefficient (K) and Scattering coefficient (S)

1 Introduction

The banana fruit is popular among different economic classes thanks to its simplicity and suitability for any event. It is cultivated in 120 nations and 86 million tons of banana fruit bunches are harvested annually around the World [29]. There are around 1000 assortments of bananas on the planet, subdivided into 50 groups [16]. Sri Lanka has a rich biodiversity of banana varieties. There are several types of bananas grown in Sri Lanka, including the Cavendish, the Red Banana, the Seeni Kesel, and the Kolikuttu Kesel. Sri Lanka boasts of 29 varieties, all of which are original to the island nation [16]. The banana (Musa sp.) is the world's fourth-largest developed organic product and the largest natural product prevalent in Sri Lanka.

Bananas are grown in almost every region of Sri Lanka, with the highest production being in the wet zone, particularly in the districts of Galle, Matara, and Ratnapura. The most commonly grown banana varieties in Sri Lanka are the Cavendish, followed by the Seeni Kesel and the Kolikuttu Kesel. The most commonly known banana is the Cavendish assortment, produced for use in business sectors [35]. Monaragala and Kurunagala districts have allocated the largest land areas, up to 6900 hectares, for the cultivation of banana plants compared to other districts. Around 2000–3000 ha of land are allocated for banana plantations in Hambantota, Kandy, Matara, Badulla, and Kegalle. Sri Lanka cultivates about 56,216 ha of banana cultivation and the cultivation produces about 44,177 bunches annually [35]. This is a strong sign that Sri Lanka has the potential to extract banana fibers for the use of its fiber and textile industry.

Banana agriculture is an important industry in Sri Lanka. The country has a long history of cultivating bananas, and the industry has expanded significantly over the years. The majority of bananas in Sri Lanka are grown by small-scale farmers, with some larger plantations also in operation. Agriculture is an essential sector of the Sri Lankan economy [20] and banana is a popular crop harvested all year on the island. Banana plant in Sri Lanka is cultivated mainly for their fruits. After the cultivation, banana plantations generate tons of agricultural waste. Without a proper practice of agricultural waste management, a vast amount of perilous waste creates serious ecological damage [22]. Agriculturists frequently face issues due to climate conditions, such as floods and substantial breezes, and these climate conditions fluctuate the value of the banana fruit. Frequently farmers utilize laborers to either cut or abandon the banana pseudostem. These are usually thrown away by the roadside. A farmer has to bear the cost of this labor and many of them deem it a wasted investment in banana fruit cultivation [13]. Moreover, dumping the banana stems adds to street pollution and the pseudostem waste that is dropped in the plantation soil transforms into organic waste, all of which contribute to environmental pollution [15]. Furthermore, pseudostem biomass is usually dumped in agricultural areas, which encourages the breeding of banana pests and causes additional environmental problems [4]. The banana pseudostem is a perilous waste in Sri Lanka since it is not put to better use. A small proportion of the waste has been utilized for a few bio-items such as paper, crafted works, pickles [13], but the majority is left as waste. The benefit of utilizing banana stems is many, from contributing to the Gross Domestic Product (GDP), of the country by producing various biodegradable items from banana wastes, to creating new job opportunities for women in the plantation sector. The income from smallholder production can support their livelihoods. Aside from these, all banana stems can be separated and changed over as a fiber, through basic machinery, enzyme, and chemical methods [25]. This can be a supportable source of income for agriculturists, as well as the development of the surrounding area as fiber manufacturing industry [10].

Sustainability is a recent concept fast gaining recognition among customers and manufacturers alike, for the advantage it brings to stakeholders. Sustainable products are not a trend; it is a need to reduce the negative impacts on the earth, society, and economy. The growing environmental awareness and preference for green fiber, textiles, and apparel motivated to development of natural fiber dyeing methods. The science of natural dyeing is the study and application of the chemical, biological, and physical processes involved in dyeing banana fibers with natural dyes. It encompasses the understanding of the properties of natural dyes, such as their chemical composition, dyeing properties, and colorfastness, as well as the techniques for extracting, preparing, and applying them to banana fiber [24].

The Sri Lankan people have been using natural dyes for about 2500 years [7]. The tradition of natural dyeing is still practiced in Sri Lanka's rural communities. Sri Lanka has a long history of natural dyes. It is a good example of Buddhist robe dyeing which is done with natural dyes that traditional event is called "Kathina civara puja". The use of natural fibers and natural dyes can help to reduce the environmental impact of the Sri Lankan textile and fiber industry. Unlike synthetic dyes, which are often derived from petrochemicals and can be toxic to the environment, natural dyes are made from renewable resources and are biodegradable. There is a growing market demand for natural and sustainable products among consumers, and the use of natural dyes can help manufacturers meet this demand and differentiate their products in the Sri Lankan market. Therefore, understanding the principles and techniques of natural dyeing can help the Sri Lankan fiber and dyeing industry produce sustainable, culturally significant, and safe products that meet the demands of today's market.

2 A Comprehensive Banana Fiber Grading System: Enhancing Quality Assessment and Market Viability

The comprehensive grading system for banana fibers aims to improve the assessment of their quality and enhance their market viability. The proposed system addresses the need for standardized criteria and procedures to evaluate banana fibers accurately. By providing a clear framework for grading, this system enables reliable and consistent classification, facilitating fair trade and promoting the utilization of natural fibers in various industries. The quality assessment and market viability of natural fibers play a crucial role in various industries such as textiles, construction, and automotive. However, the lack of standardized criteria and evaluation methods often hinders accurate and consistent grading. Therefore, this study aims to address this gap by proposing a systematic and reliable grading system specifically tailored to banana fibers. By providing a clear framework and guidelines, this system aims to improve the overall quality assessment process and facilitate fair trade practices. Based on the obtained results, a novel grading system for different banana fiber varieties, specifically the Sri Lankan cultivar, was introduced. This grading system incorporates five parameters: strength, fineness or coarseness, color, luster, and contamination of impurities. Corresponding to these parameters, four grading points were established: one (1), two (2), three (3), and four (4). This system provides a standardized approach for assessing and categorizing banana fibers, aiding in quality control and market evaluation [2].

2.1 Properties of Quality Graded Sri Lankan Banana Varieties: The High-Quality Banana Fiber Variety: Grading Points 1

The high-quality graded Sri Lankan banana varieties exhibit several notable properties. These properties contribute to their high quality and desirability for various applications. Some key properties of the first-grade Sri Lankan banana varieties include:

Color: The high-quality graded Sri Lankan banana varieties typically exhibit vibrant and appealing colors. They may range from golden yellow to other attractive shades, enhancing their visual appeal.

Luster: These banana varieties often possess a noticeable luster or shine on their surface. This luster adds to their overall appearance and can contribute to the visual quality of products made from these fibers.

Softness: The fibers of high-quality graded Sri Lankan banana varieties tend to be relatively soft in texture. This softness provides a pleasant tactile experience and can contribute to the comfort of products made from these fibers.

Strength: The high-quality graded Sri Lankan banana varieties are known for their strength. They exhibit good tensile strength, enabling them to withstand pulling forces and resist breaking or deformation under stress.

Fineness: These banana fibers are generally fine in nature, with a smaller diameter. This fineness contributes to their smoothness and ability to be woven or processed into various textile products.

Low Contamination: The high-quality graded Sri Lankan banana varieties exhibit low contamination of impurities. They are relatively clean and pure, making them suitable for applications where purity is important.

These properties make the high-quality graded-grade Sri Lankan banana varieties highly suitable for a range of applications, including textiles, crafts, and other industries. Their vibrant colors, luster, softness, strength, fineness, and low contamination contribute to their overall quality and make them a desirable choice for various end products [2].

2.2 The Medium-Quality Banana Fiber Variety: Grading Points 2

The medium-quality banana fiber variety exhibits the category of medium fine, suggesting a diameter within the range of $\leq$ 85–87 μm. This range signifies a relatively thicker texture compared to finer fibers. The presence of medium contamination of impurities implies the existence of a moderate level of foreign substances within the fibers, potentially affecting their purity and overall quality.

Additionally, the tensile strength of the fibers falls within the range of $\geq$ 2.4–1 N, suggesting their ability to withstand pulling forces to varying degrees. This range signifies a moderate tensile strength, which may be appropriate for certain applications but may have limitations in terms of durability and resistance to stress [2].

2.3 *Ordinary Banana Fiber Variety: Grading Points 3*

The ordinary banana fibers are categorized within the medium coarser category, suggesting a diameter range of $\leq$ 87–110 μm. This range signifies a relatively thicker texture compared to finer fibers, which may impact their softness and overall handling characteristics. The tensile strength of the fibers falls within the range of $\geq$ 2.4–1 N, suggesting their ability to withstand pulling forces to varying degrees. This range signifies a moderate tensile strength, which may be sufficient for certain applications but may have limitations in terms of durability and resistance to stress [2].

2.4 *Low-Quality Banana Fiber Variety: Grading Points 4*

The low-quality banana fibers show a coarser category, with a diameter range of $\leq$ 111 μm. This range signifies a thicker texture compared to finer fibers, which may negatively impact their softness and overall handling properties. Moreover, the fibers have high contamination of impurities, indicating a significant presence of foreign substances within the fibers, which can negatively affect their purity and overall quality. The tensile strength of the fibers falls within a range of $\leq$ 1 N, indicating a low level of strength. This suggests that the fibers have limited resistance to pulling forces and may be more prone to breakage or deformation. The color of banana fibers shows dark shades [2] (Table 1).

Table 1 Properties of quality-graded Sri Lankan banana varieties

Quality of banana fibers	Grading points	Type of Sri Lankan banana variety
High-quality banana fiber variety	1	Ambun (Genome type AAA, Cavendish group) and Kolikuttu (Genome type AB, Kolikuttu group)
Medium-quality banana fiber variety	2	Atamuru (Genome type ABB, plantain varieties, or cooking varieties) and Puwalu (Genome type AB, Kolikuttu group)
Ordinary-quality banana fiber variety	3	Seeni (Genome type AAB, Mysore group), Suwandel (Genome type AB, Kolikuttu group), and Ambul (Genome type AAB, Mysore group)
Low-quality banana fiber variety	4	Rath kehel (Genome type AB, Kolikuttu group), Anamalu (Genome type AAA, Cavendish group), and Alu kehel (Genome type ABB, plantain varieties)

3 From Trash to Treasure: Sustainable Utilization of Banana Fibers

Banana fiber production has gained significant attention in recent years as a sustainable and eco-friendly alternative to conventional textile materials and handicrafts. Banana fiber has the potential for industrial applications in the manufacturing of paper, rope, mat industries and the production of other composite materials for vehicles [22]. Soni & Patel [31] found that banana fiber can be used to produce car mats by mixing a yarn with jute, cotton, and banana fiber to make a nonwoven fabric. Banana fibers are a cellulose substitute for high-quality paper-making and reinforced composites [25]. Mohiuddin et al., [19] Highlight how the Philippines, Malaysia, Japan, and Korean countries use banana fiber as blended material in their textile industry. European countries were found to make socks and gloves using banana fibers [19]. Mohapatra et al. [18] mentioned that banana fibers can be used as environmentally degradable bags (eco-bag), hard paper, and cardboard. Sri Lanka, known for its abundant banana cultivation, holds great potential for the development of a sustainable banana fiber industry. In Sri Lanka, banana fibers can be used as bags, mates, crochet bags, hats, mats, packaging bags, earring, and jewelry as shown in Fig. 1.

4 Optimizing Reactive Dyeing Recipe and Conditions for Banana Fiber: Enzyme Treatment and Scoured-Bleached Fibers

4.1 Enzyme and Chemical-Treated Ambun Banana Fibers

The raw banana fibers were treated with enzyme and 6% H_2O_2, 2% Na_2SiO_3, 3% NaOH. Treatment of enzyme and 6% H_2O_2, 2% Na_2SiO_3, and 3% NaOH, combined treatment has reduced diameter of 48.8 μm and a strength value is 2.931 N. Enzyme and chemically combined treatment has reduced the diameter from 140 to 48.86 μm which is about 71% diameter reduction compared to mechanically extracted fiber diameter value. From the experimental results, the raw banana fibers have been improved their fineness by approximately 71%. Treatment of enzyme and 7% H_2O_2, 2% Na_2SiO_3, and 2.5% NaOH, combined treatment has reduced diameter of 49.7 μm and strength value reduced by 1.01 N. These treated fibers' samples break easily. The results indicate that the diameter and single fiber strength of fibers are affected after the enzyme and chemical 114 μm combined treatments. Pectinase enzyme-treated and chemically treated banana fibers are shown in Fig. 2. This treatment removes the lignin impurities and reduces the impurities that result from the reduction of diameter [2].

Fig. 1 Handicrafts from banana fibers

4.2 Consumer Perception of Natural Fiber Synthetic Dyes

Consumers usually judge the aesthetic appeal of products created from banana fiber. This involves assessing the hue, texture, and general appearance of anything created from banana fiber, such as clothes, accessories, and handicrafts. Scoured and bleached banana fibers were applied with the reactive class of dye and it was found that the dye absorption properties and the wash fastness properties satisfied the quality level (5 is excellent) required for textile fiber. A reactive class of dyes was used to dye banana

Fig. 2 Enzyme and chemical-treated fibers

fibers because of their superior performances in dyeing cellulosic goods. Figure 3 shows the reactive dyeing mechanism of banana fiber.

Synthetic dyes have been widely used in the textile industry for many years due to their ease of use, cost-effectiveness, and ability to produce a wide range of colors. However, there are several disadvantages associated with their use, including environmental impact, health hazards, lack of biodegradability, and limited color fastness. By contrast, natural dyes offer a more sustainable and healthy alternative to synthetic dyes, as they are often derived from renewable resources and are biodegradable. In addition, natural dyes can produce unique and varied color palettes that cannot be replicated by synthetic dyes.

Synthetic dyes and natural dyes are two distinct categories of colorants used in the textile industry, each with its own advantages and limitations. When it comes to dyeing banana fiber, the choice between synthetic dyes and natural dyes becomes a crucial decision, as it can significantly impact the final product's quality, sustainability, and market appeal.

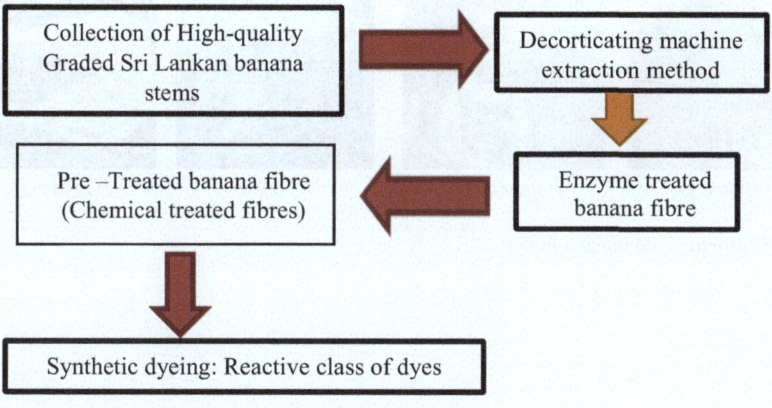

Fig. 3 Process flow in the dyeing mechanism of banana fiber

In contrast, natural dyes are derived from plant, animal, or mineral sources and have been used for centuries in traditional textile dyeing practices. Natural dyes offer a unique color palette, often characterized by earthy tones and subtle variations. They are perceived as environmentally friendly and sustainable since they are obtained from renewable resources and tend to biodegrade more readily than synthetic dyes. However, natural dyes may pose challenges in terms of color fastness, limited color range, and scalability for large-scale production. The choice between synthetic dyes and natural dyes for dyeing banana fiber involves considering various factors, including desired color range, colorfastness requirements, sustainability goals, and market preferences. Understanding the advantages and limitations of each dyeing approach can help inform decisions regarding color selection, sustainability considerations, and the overall quality of banana fiber-based products.

The exhaust method under the reactive dyeing of banana fiber shows that these fibers had better color strength and more luster. The reactive dyeing process of the banana fiber is darker and less saturated. The results from the reflectance curve established that the dye absorption tendency of dyed banana fiber is similar to that of cotton-dyed fiber. The single-fiber strength of pre-treated fibers is comparable to that of dyed fibers. The average value of enzyme, 6% H_2O_2, 2% Na_2SiO_3, 3% NaOH, combined treated fiber, and dyed fiber is 2.931 N. The diameter of dyed fibers was raised from 48.86 to 80.11 m, a 39% increase over pre-treated fibers. Banana fibers produced visually darker blue color shades when compared to cotton fibers when dyeing fibers using a similar recipe of primary three colors such as red, blue, and yellow were applied on both cotton and banana fiber. Banana fiber material can achieve the desired shade with less dye in the dye liquid than cotton material. Figure 4 shows the dyed banana fiber samples. However, cotton fibers instead of banana fibers produced vibrant colors and shiny effects. The results of the washing fastness of the dyed banana fiber varied with the types of colors. However, the results indicate "good to excellent" washing fastness of dyed banana fiber.

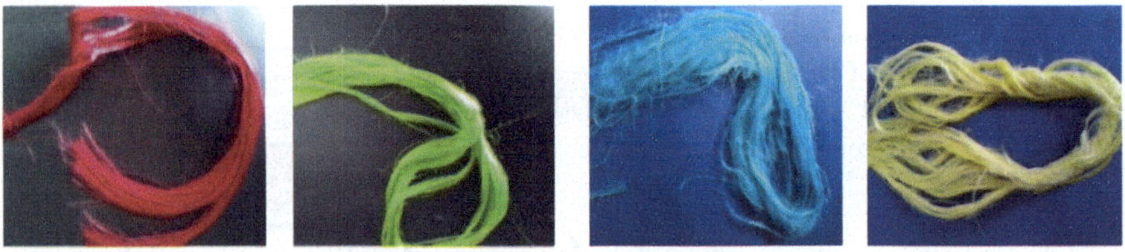

Figure a *Figure b* *Figure c* *Figure d*

Fig. 4 Synthetic dyed banana fibers

4.3 Reactive Dyeing Method for Cotton and Banana Fiber Blend Fabric: Achieving Vibrant and Sustainable Coloration

The combination of cotton and banana fibers offers a sustainable method for producing textiles by fusing the appealing qualities of cotton with the unique attributes of banana fiber. Reactive dyes can be used in the dyeing process to produce brilliant, long-lasting coloration while assuring compatibility with both fiber types. Reactive dyes are well known for their propensity to establish covalent connections with cellulose fibers, which makes them ideal for fabrics made from a mixture of banana and cotton fibers. This dyeing technique produces vibrant and long-lasting hues on the cloth surface due to outstanding color absorption and retention capabilities. Additionally, reactive dyes have benefits such as strong lightfastness and washability, ensuring color stability. Figure 5 shows synthetic-dyed banana and cotton-blended fabrics, and Fig. 6 shows cotton and banana-blended garments.

5 Eco-Friendly Natural Dyeing of Banana Fiber

Consumers are becoming more conscious about the environmental effects of the things they buy as eco-awareness rises. Due to its renewable source, banana fiber is frequently regarded as sustainable dyeing, which can be a big marketing feature. It is critical to evaluate how consumers view eco-friendliness and sustainability in items made using banana fiber.

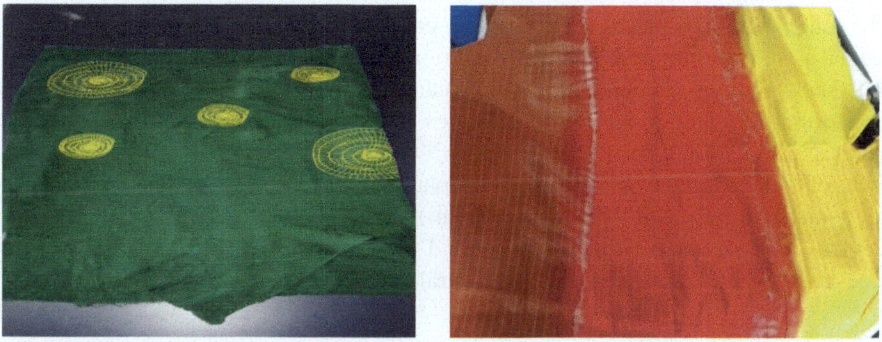

A B

Fig. 5 Synthetic-dyed banana and cotton-blended fabrics

A B C

Fig. 6 Synthetic-dyed banana and cotton-blended garments

5.1 Sources of Natural Dyes

The sources of natural dyes can be broadly categorized into three types: plant-based, animal-based, and mineral-based. In addition to these primary sources, other materials such as fungi, lichens, and insects can also be used to extract natural dyes. The availability and quality of natural dyes can vary depending on factors such as geographic location, climate, and season, which can also influence the color and properties of the resulting dyes. In this chapter, plant-based banana pseudostem sap (Musa acuminata), jamun fruit (Syzygium cumini), coconut shells [Cocos nucifera (L.)], Ceylon green spinach (*Basella alba*), and Venivelgeta/Tree Turmeri (Coscinium fenestratum) natural sources involved natural dyeing methods are mainly considered.

5.2 Natural Dye Sources and Their Properties for Banana Fibers: Exploring Sustainable Coloration Options

Across the world, synthetic dyes are widely used. More than 100,000 commercial dyes in a wide spectrum of hues are available on the market. The annual production of dye is a remarkable 700,000 tons [12]. It is concerning to learn about the harmful effects of the dyes in textile industry effluents. These dyes are not only toxic but also resistant to microbial breakdown, which makes them even more dangerous. The fact that they can create probable carcinogens is alarming. Moreover, the bioaccumulation and biomagnification effects are also worrying as they can increase the concentration of pollutants in organisms and food chains. We need to take measures to reduce the use of these harmful dyes and find alternative, eco-friendly options [5].

"Using natural dyes" instead of synthetic ones can boost the demand for natural dye extraction. Growing environmental awareness and a preference for green textile concepts are recognized by both customers and manufacturers, with their own viewpoints based on the present and future demand market [30]. Banana fibers, derived from the pseudostem of the banana plant, possess inherent strength and eco-friendliness, making them an ideal choice for sustainable textile applications. By harnessing the vibrant pigments present in natural dye sources, such as the sap of banana pseudostem, jamun fruit, coconut shells, Ceylon green spinach, and Venivelgeta/Tree Turmeric, unique and organic color variations can be achieved on banana fibers. One of the key aspects explored in this chapter is the mordant-free dyeing technique for banana fibers. Mordants, chemical substances used to enhance dye absorption and colorfastness, are often avoided in sustainable dyeing practices. The chapter discusses the dyeing process without the use of mordants, focusing on optimizing dye bath conditions, such as pH adjustment and temperature, to enhance dye penetration and color retention on banana fibers.

This chapter focuses on the utilization of natural dye sources, including banana pseudostem sap (*Musa acuminata*), jamun fruit (*Syzygium cumini*), coconut shells [*Cocos nucifera* (L.)], Ceylon green spinach (*Basella alba*), and Venivelgeta/Tree Turmeric (*Coscinium fenestratum*), for dyeing banana fibers without the use of mordants. Natural dyeing techniques offer sustainable and environmentally friendly alternatives to synthetic dyes, and exploring these diverse dye sources enhances the color palette and aesthetic appeal of banana fiber textiles.

5.2.1 Banana Pseudostem Sap (*Musa Acuminata*)

As a more environmentally friendly replacement for metallic mordants, biomordants are gaining popularity. The interactions between fiber, mordant, and dye in addition to the chemical composition of mordants significantly affect the fastness characteristics of naturally colored fibers [23]. Banana pseudostem sap is a significant natural resource with numerous existing and prospective future uses. For natural dyes, banana pseudostem sap is utilized as a biomordant [26]. The chemical composition of banana pseudostem sap is anthocyanins, alkaloids, and phenolic compounds. The chemical composition of banana sap makes it highly soluble when linked to the fabric, making it difficult to remove from clothing even after many washes. Tannins are organic pigments with a brown color which are found in banana sap. In batik dyeing, the color brown is frequently applied [21]. A hot water-based extraction method was used for the extraction of Banana pseudostem. Banana pseudostem sap is emerging as a

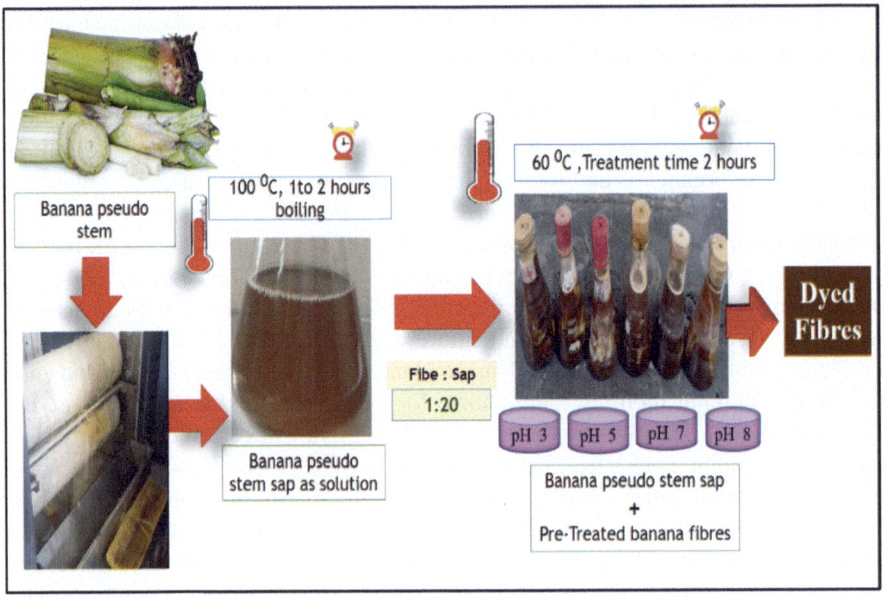

Fig. 7 Banana pseudostem sap (*Musa acuminata*) extraction and dyeing method

promising natural dye source for textiles, producing a wide range of vibrant colors [3]. The banana pseudo stem was cut off and the banana (*Musa sapientum)* stem was cleaned. Fresh stems were sliced after being divided into one-meter lengths. A roller calendar machine was used to remove the sap from the stem. Filtered solutions were made from the sap combination and kept in see-through plastic bottles. To prevent evaporation and potential reactions, the sap was maintained in a cold environment and shielded from the sun [26]. Post-mordanting was carried out according to the exhaust method. pH banana sap juice ranges from 4.5 to 5.5. The optimal dyeing period for banana fiber coloration with banana floral stem sap is 1 to 2 h, producing a slightly deeper hue as the dyeing duration increases. With the application of these banana sap solutions, excellent color fastness qualities also appeared for 100 °C [26]. Banana pseudo stem sap produces a brown tint and tone colors. Banana sap can be used for sustainable coloring, and while using this green ingredient, zero wastewater discharge can be achieved. The banana sap extraction and process flow of dyeing mechanism are illustrated in Fig. 7.

5.2.2 Banana Pseudostem Sap (*Musa Acuminata*) and Jamun Fruit (*Syzygium Cumini*)

The use of natural dyes for staining different biological tissues will lower the cost of purchasing synthetic dyes and reduce their negative impact on people and the environment. Jamun fruit juice stains on textiles are caused by the chemical or physical

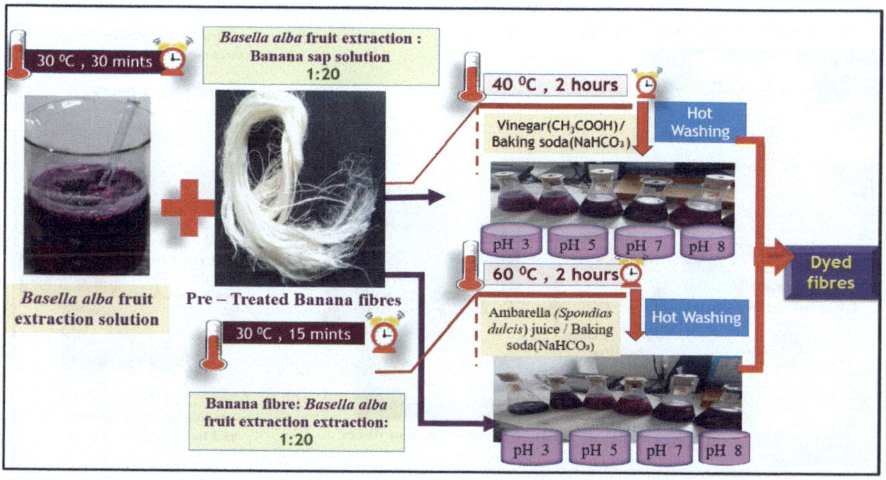

Fig. 8 Banana pseudostem sap (*Musa acuminata*) and jamun fruit (*Syzygium cumini*) extraction and dyeing method

interaction of materials, leaving permanent marks that can be difficult to remove. *Syzygium cumini* (L.) is commonly known as jamun fruit and it belongs to the family of Myrtaceae [32]. Fresh jamun fruit (*Syzygium cumini*) can be extracted from violet dye and mixed with banana sap solution at a 4:1 ratio. The pH of the jamun fruit extraction was 3.77. The pH of the dye bath increased from 2 to 5 improving the color strength and dyeability. Since the pH was more than 7, the K/S values of dyeing techniques dramatically decreased. Banana stems sap with the jamun fruit (*Syzygium cumini*) dyeing procedure showed the greatest improvement in terms of color fastness properties. Jamun fruit (*Syzygium cumini*) dyes are more effective at a temperature of 90^0 C. Jamun fruit (*Syzygium cumini*) dyes produce purple tint colors. The dyed fiber is more environmentally friendly and sustainable as a result of the use of bio-mordants, which produce great color strength and superior fastness feature. The jamun fruit (*Syzygium cumini*) extraction and process flow of dyeing mechanism are illustrated in Fig. 8.

5.2.3 Banana Pseudostem Sap (*Musa Acuminata*) and Coconut Shells [*Cocos Nucifera* (L.)]

In Sri Lanka, banana cultivation and coconut shells generate large quantities of biomass. The pseudostem and peduncle of banana crops contain a significant amount of natural fiber. *Cocos nucifera* (L.) belongs to the family of Arecaceae. The tough coconut "shell" is made up of the exocarp, which is the shiny outside skin. In order to assess the dyeing and biomordanting properties of coconut shell extraction, several research investigations were conducted [33]. Phytochemical constituents of coconut shell extraction contain Saponins, Phenols, Tannins, Flavonoids, and Glycosides. A

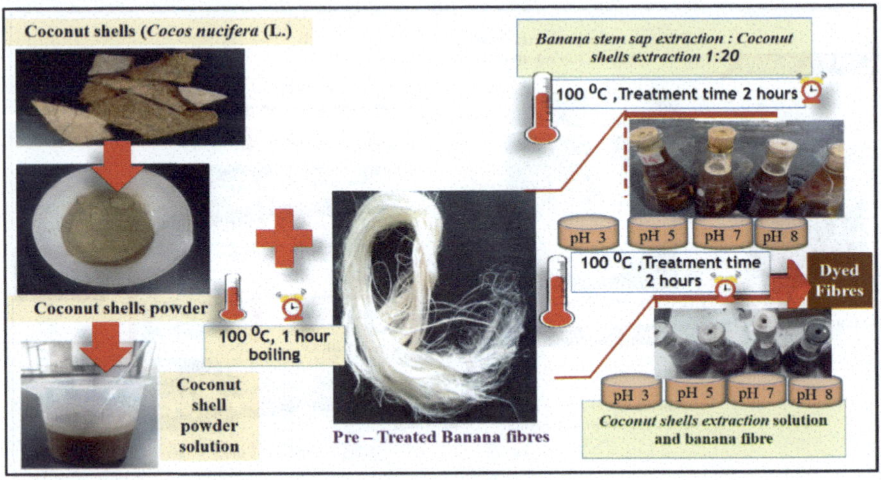

Fig. 9 Banana pseudostem sap (*Musa acuminata*) and coconut shells [*Cocos nucifera* (L.)] extraction and dyeing method

coconut shell extract from an alkaline pH was utilized to give cotton (Cocos nucifera Linn) flame-retardant and thermal stability, and it was shown that this increased the oxygen index value by 72.2% [33]. Coconut shells and banana pseudostems sap can effectively extract the colorants using the hot water extraction technique. Mordants are substances that help bind natural dyes to textile fibers in many cases [1]. The fiber of bananas is mordanted with vinegar and banana pseudostem sap. To create a mordant bath, the appropriate amount of mordant is dissolved in water. Brown colors were then obtained in light shades in vinegar and deep shades in banana pseudostem sap. Coconut shells' dyes produce brown tint and tone colors. The color fastness to hand wash and light fastness of the naturally dyed samples were evaluated and it was found that the samples showed good color fastness with a rating of 4–5 on the scale. A significant step toward environmentally friendly fiber dyeing techniques is the investigation of banana pseudostem sap and coconut shells as natural dye supplies for banana fibers. They are beneficial contributions to the fiber industry's efforts to achieve sustainability because of their eco-friendly qualities, a wide variety of color options, and potential for waste reduction. The coconut shells [*Cocos nucifera* (L.)] dye extraction and process flow of dyeing mechanism are illustrated in Fig. 9.

5.2.4 Banana Pseudostem Sap (*Musa acuminata*) and *Basella Alba*

The fruit of *Basella alba* is not commonly consumed or used for culinary purposes and is often considered waste because of its relatively low market value and limited applications. As a result, many people tend to discard the fruit, and it is not commonly utilized in the food industry or for other commercial purposes [13]. Green spinach can maintain the sustainability of the natural dyeing process, which can lead to profitability for Sri Lanka. *Basella alba* belongs to the family of Basellaceae [13]. Green spinach (*Basella alba*), red spinach (*Basella rubra*), and tree spinach (*Talinum fruticosum*) are three types of spinach consumed in Sri Lanka [7]. *Basella alba* fruits are round, spherical, meaty, and purple in color [6]. This *basella alba* fruit contains betacyanin and flavonoid pigments [17, 27]. Fruit extraction juice from *Basella alba* can be used as a natural dye for textiles since it contains pigments [17]. Juice from *Basella alba* fruit is used to create a bright color that changes as the pH balance between acid and base changes [6]. By combining several mordants, cotton could be successfully dyed with *Basella alba* fruit extract to produce a variety of light color [17]. This natural extracted dye has a pH of 4 and is purple in color. The color strength (K/S) values of banana fiber dyed with banana sap and *basella alba* fruit showed that strength values were higher at 40 °C compared to 60 °C. Raising the pH and temperature can damage the betacyanin molecule. The naturally extracted dye color is purple, and this dye applied to different pH levels produces different colors on banana fibers with acceptable fastness properties. To maintain the proper pH during the dyeing process, baking soda ($NaHCO_3$) and ambarella (*Spondias dulcis*) juice and vinegar were used as pH regulators. Astringency and terpene-like flavors can be found in the acidic fruit ambarella [33]. *Basella alba* produced purple tint at pH 3, yellow tint at pH 5, and pink tint at pH 8 which are important findings. Banana fiber dyed with *Basella alba* extraction dye found that vinegar and baking soda-treated fibers imparted good color strength and fastness properties. The *Basella alba* fruit dye extraction and process flow of the dyeing mechanism are illustrated in Fig. 10.

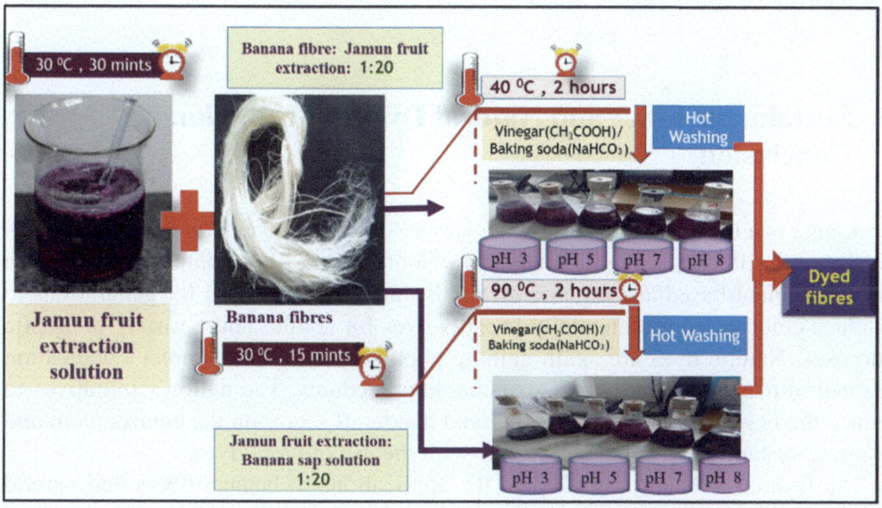

Fig. 10 Banana pseudostem sap (*Musa acuminata*) and *Basella alba* extraction and dyeing method

5.2.5 Banana Pseudostem Sap (*Musa Acuminata*) and Venivelgeta/Tree Turmeric (*Coscinium Fenestratum*)

Tree Turmeric is found naturally in Sri Lanka's low region (elevation 500 m), wet zone (annual rainfall > 2500 mm), which is known as venial-geta. Under the names False Calumba and Tree Turmeric, the yellow dye was taken from the stem and shipped to Europe [9]. Tree Turmeric is beneficial for inflammations, wounds, ulcers, jaundice, burns, skin conditions, stomach illnesses, diabetes, fever, general debility, and as a blood cleanser [3].

The bright yellow and orange hues of Venivelgeta are caused by the presence of berberine alkaloids and tannins. Tannins help the dye adhere to banana fibers [8]. Tree Turmeric can effectively extract the colorants using the hot water extraction technique. Light yellows to deep yellow are available in a range of colors due to Venivelgeta-based dyes. Natural dyes made from Venivelgeta have good light and wash fastness, especially when used with and without mordants. This dye extraction made from venivelgeta exhibits outstanding lightfastness and wash resistance, especially when used with and without mordants [11]. pH of Tree Turmeric (*Coscinium fenestratum)* was 7. In the effort of sustainable and historically established methods for dyeing banana fiber with this Tree Turmeric, its eco-friendliness, colorfastness, and cultural importance make it an invaluable colorant. There have been numerous studies on turmeric powders (*Curcuma longa* L), but there are few specifics on the techniques used to color trees turmeric (*Coscinium fenestratum*) [28]. The Tree Turmeric (*Coscinium fenestratum*) dye extraction and process flow of the dyeing mechanism are illustrated in Fig. 11.

6 Sustainable Fiber and Natural Dyes in Sri Lanka: Conclusion

Sri Lanka is a unique source of a wide variety of natural dyes due to its vast biodiversity and different climatic zones. This South Asian island nation is home to a wealth of plant-based and organic materials that have been used for generations to produce colorful, environmentally friendly dyes for textiles and a variety of artistic purposes. Natural dyes are again gaining popularity due to Sri Lanka's dedication to environmentally friendly and sustainable procedures. The nation's initiatives to reduce the negative effects of the fiber and handcraft sector on the environment and promote sustainable farming are in line with the use of these dyes.

The fashion industry has adopted the applications of banana fibers and natural dye methods to get a competitive advantage in the market. Sustainable natural fibers are increasing in popularity and value among handcraft, textile, and apparel manufacturers due to recent global campaigns that focus on environmental safety when manufacturing garments. The end consumer must be engaged in order to develop and maintain an affinity for natural fibers. Naturally, dyed banana fibers are not only

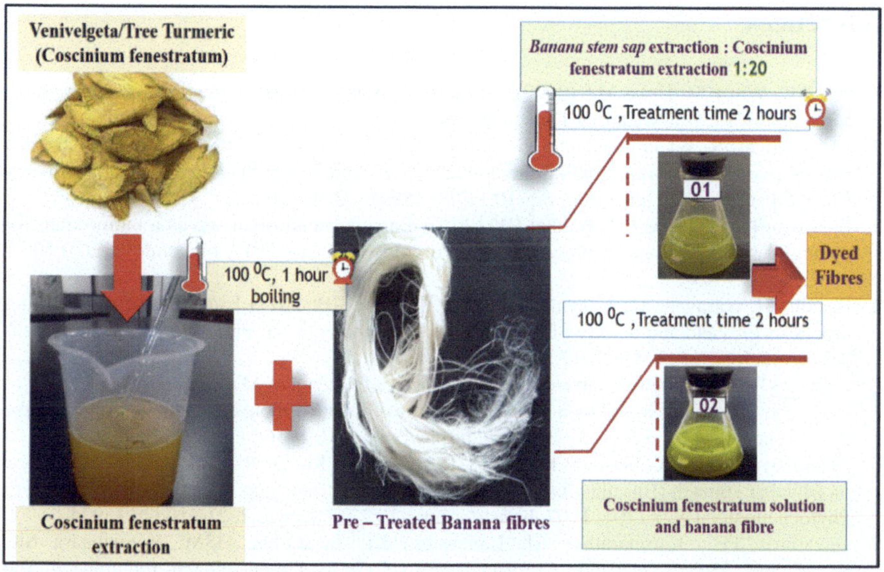

Fig. 11 Banana pseudostem sap (*Musa acuminata*)and Venivelgeta/Tree Turmeric (*Coscinium fenestratum*) extraction and dyeing method.

integral to the fiber and craft industry but play an important role in economic development. The banana fiber decorticating machine extraction can be promoted among poor farmers and small rural entrepreneurs. Researchers have recently concentrated on developing the usage of banana fibers as a natural raw material. This has lead researchers to acquire enough practical knowledge on extracting banana fibers and natural dyeing methods efficiently. They have found that these methods add value, decrease the cost of extraction, select suitable variety and dyeing, and motivate the large-scale use of these fibers. Advanced technology needs to be developed in order to increase fiber yields from extraction processes and low water-consumption dyeing methods. Sri Lanka offers unique natural dye sources for banana fiber such as Banana pseudostem sap (*Musa acuminata*) extraction, jamun fruit (*Syzygium cumini*) extraction, coconut shells powder (*Cocos nucifera*), Ceylon green spinach (Basella alba) extraction, and Tree Turmeri (*Coscinium fenestratum*) powder. The natural shades obtained from these dyes are attractive and can add value to banana fiber-made products.

In conclusion, Sri Lanka is a great source of natural dyes due to its abundance of biodiversity and climate variability. These dyes continue to play a significant role in Sri Lankan culture and contribute to the country's sustainable and environmentally friendly approach to arts and craftsmanship, from traditional textile crafts to present-day eco-conscious activities.

References

1. Aye TK, Win KM, Htwe MT (2000) Application of natural dyes on textiles producing from some plants. In: 2nd Myanmar Korea Conference Research Journal, pp 511–520
2. Balakrishnan S, Wickramasinghe G, Wijayapala US (2020) A novel approach for banana (Musa) pseudo-stem fibre grading method: extracted fibres from Sri Lankan banana cultivars J Eng Fibers Fabr 15. https://doi.org/10.1177/1558925020971766
3. Barahapurkar S, Purwar R, Baldua RK (2017) Banana pseudostem sap as a biomordant for dyeing of silk with Celosia Flower. Fibers Polym 21(9):2010–2017. https://doi.org/10.1007/s12221-020-9045-2
4. Brindha R, Narayana C, Vijayalakshmi VJ, Nachane RP (2017) Effect of different retting processes on yield and quality of banana pseudostem fiber. J Nat Fibers 16(1):58–67. https://doi.org/10.1080/15440478.2017.1401505
5. Carneiro PA, De Aragão Umbuzeiro G, De Oliveira DP, Zanoni MVB (2010) Assessment of water contamination caused by a mutagenic textile effluent/dyehouse effluent bearing disperse dyes. J Hazard Mater 174(1–3):694–699. https://doi.org/10.1016/j.jhazmat.2009.09.106
6. Chaurasiya A, Pal RK, Verma P, Katiyar A, Razauddin N, Kumar N (2021) An updated review on Malabar spinach (Basella alba and Basella rubra) and their importance. J Pharmacognosy Phytochemistry 10(2):1201–1207. https://doi.org/10.22271/phyto.2021.v10.i2p.13974
7. Gayathree THI, Karunarathne SI, Ranaweera LT, Jayarathne HSM, Kannangara SK, Ranathunga APDT, Weebadde CP, Ranaweera LT (2020) Green-spinach, red-spinach, and tree-spinach ('Three-Fold Spinach' in Sri Lanka): an insight into phylogenetics and consumer preference. Emirates J Food Agric. https://doi.org/10.9755/ejfa.2020.v32.i2.2065
8. Hosen MD, Rabbi MF, Raihan MA, Mamun MA (2021) Effect of turmeric dye and biomordants on knitted cotton fabric coloration: a promising alternative to metallic mordanting. Cleaner Eng Technol 3:100124. https://doi.org/10.1016/j.clet.2021.100124
9. Jayaweera DMA (2006) Medicinal Plants (Indigenous and Exotic) used in Ceylon. Nat Sci Found Sri Lanka, Colombo, Sri Lanka 4:70–71
10. Jideani AIO, Anyasi TA (2020) Banana nutrition: function and processing kinetics. BoD—Books on Demand
11. Kashyap S, Kapoor N, Kale RD (2016) Coscinium fenestratum: callus and suspension Cell culture of the endangered medicinal plant using Ver-micompost extract and coelomic fluid as plant tissue culture media. Am J Plant Sci 2016(April):899–906
12. Katheresan V, Kansedo J, Lau SY (2018) Efficiency of various recent wastewater dye removal methods: a review. J Environ Chem Eng 6(4):4676–4697. https://doi.org/10.1016/j.jece.2018.06.060
13. Kumar S, Prasad AK, Iyer SV, Vaidya SK (2013) Systematic phar-macognostical, phytochemical and pharmacological review on an ethno medici-nal plant, Basella alba L. J Pharmacognosy Phytotherapy 5(4):53–58. https://doi.org/10.5897/jpp12.0256
14. Kumar MKD (2011) Comparative study of pulping of banana stem. Int J Fibre Text Res 1(1):1–5
15. Kun Li, Shiyu Fu, Huaiyu Zhan, Yao Zhan LAL (2010) Analysis of the chemical composition and morphological structure of banana pseudo-stem. BioResources 5(2):576–585
16. Liyanage A, Manawaprema M, Mendis M (1998) Differentiation of A & B genome of banana and plantain (Musa spp.) by esterase enzyme. J Natl Sci Found Sri Lanka 26(2):125. https://doi.org/10.4038/jnsfsr.v26i2.3560
17. Mitra A, Das SK (2015) Fabric dyeing with natural dye extracted from Basella alba fruit and spectroscopic analysis of the extract at different conditions. J Chem Pharm Res 7(12):1117–1124
18. Mohapatra D, Mishra S, Sutar N (2010) Banana and its by-product utilization: an overview. J Sci Ind Res 69:323–329
19. Mohiuddin AKM, Saha ML, Hossian S, Ferdoushi A (2014) Usefulness of banana (*Musa paradisiaca*) wastes in manufacturing of bio-products: a review. Agriculturists 12(1):148–158. https://doi.org/10.3329/agric.v12i1.19870

20. Nanayakkara MPA, Pabasara W, Samarasekara A, Amarasinghe D, Karunanayake L (2018) Extraction and characterisation of cellulose materials from Sri Lankan agricultural waste. In: Proceedings of the international forestry and enviroment symposium. https://doi.org/10.31357/fesympo.v22i0.3459
21. Nuriana W, Winarni M, Suryono S (2019) Utilization of sap from part of Kepok banana tree (Musa Mcuninata Balbisianacolla) with variation of extraction solutions as textile dyes. J Phys. https://doi.org/10.1088/1742-6596/1381/1/012002
22. Padam BS, Tin HS, Chye FY, Abdullah MI (2014) Banana by-products: an under-utilized renewable food biomass with great potential. J Food Sci Technol 51(12):3527–3545. https://doi.org/10.1007/s13197-012-0861-2
23. Periyasamy AP (2022) Natural dyeing of cellulose fibers using Syzygium cumini fruit extracts and a bio-mordant: a step toward sustainable dyeing. Sustain Mater Technol 33:e00472. https://doi.org/10.1016/j.susmat.2022.e00472
24. Rajeshkumar G, Devnani G, Sinha S, Sanjay, Siengchin S (2022) Bast Fibers and their composites: processing, properties and applications. Springer, Berlin
25. Ray DP, Bhaduri SK, Nayak LK, Ammayappan L, Manna K, Das K (2012) Utilization and value addition of banana fibre—a review. Agric Res Commun Centre 33(1):46–53
26. Repon MR, Mamun MA, Islam MT (2016) Eco-friendly cotton coloration using banana (Musa sapientum) waste: optimization of dyeing temperature. Univ J Eng Sci 4(1):14–20. https://doi.org/10.13189/ujes.2016.040102
27. Reshmi SK, Aravindhan KM, Devi PS (2012) The effect of light, temperature, pH on stability of betacyanin pigments in Basella alba fruit. Asian J Pharm Clin Res 5(4):107–110
28. Sachan K, Kapoor VP (2007) Optimization of extraction and dyeing conditions for traditional turmeric dye. Indian J Tradit Knowl 6:270–278
29. Senthilarasi M, S M. M. R., Maheesa M, Sivaranajani R (2020) Deep learn bananas: a transfer learning for banana variety classification. In: Springer eBooks 569–577). https://doi.org/10.1007/978-981-15-8697-2_53
30. Shivankar V, Mukhopadhyay S (2019) Some studies on 100% banana parallel laid and 60–40% banana: polypropylene cross laid non-woven fabrics. Fashion Text 6(1). https://doi.org/10.1186/s40691-018-0157-x
31. Soni M, Patel S (2016) Application of banana nonwoven fabric for car mats 4(03):88–90
32. Suabjakyong P, Romratanapun S, Thitipramote N (2014) Extraction of natural histological dye from black plum fruit syzygium extraction of natural histological dye from black plum fruit (Syzygium cumini) J Microsc Soc Thai 4:13–15
33. Teli MD, Pandit P (2018) Development of thermally stable and hygienic colored cotton fabric made by treatment with natural coconut shell extract. J Ind Text 48(1):87–118. https://doi.org/10.1177/1528083717725113
34. Tushar KV, George S, Remashree AB, Balachandran I (2008) Coscinium fenestratum (Gaertn.) Colebr—a review on this rare, critically endangered and highly-traded medicinal species J Plant Sci 3:133–145
35. Wasala W, Dharmasena DAN, Dissanayake T, Thilakarathne, BMKS (2014) Physical and mechanical properties of three commercially grown banana (*Musa acuminate* Colla) cultivars in Sri Lanka. Trop Agric Res 24(1):42. https://doi.org/10.4038/tar.v24i1.7988

Role of Traditional Crafts in Sustainable Development: Reflections on Dumbara Weaving

Sulari De Silva

Abstract This study examines how traditional crafts intersect with sustainable development, aiming to uncover factors that support the ongoing viability of rural artisanal livelihoods. The study centers on Dumbara weaving, Sri Lanka's ancient textile tradition, to explore its role in fostering long-term sustainability. For approximately six centuries, Dumbara weaving has been upheld by a few artisan families residing in the village of Talagune nestled in the central hills of Sri Lanka. The Dumbara artisan community, historically relegated to a low caste status, continues to endure marginalization and oppression to this day. The present study proposes that the endurance of Dumbara weaving over time is attributed to its integration into social institutions such as familial and kinship networks, alongside non-kin artisan-patron relationships. Artisans utilize both forms of social relationships to mitigate their marginalized status, perceiving their craft as more than just a means of livelihood. The central focus of this study is on how the Dumbara artisan community negotiate the complexities of tradition and innovation, authenticity and commercialization within a socio-economic structure where elite patrons exert substantial influence. By exploring this two-way relationship, the study deals with how Dumbara weaving contributes to the attainment of Sustainable Development Goals (SDGs), particularly emphasizing the sixteenth goal outlined by the United Nations. This goal underscores the significance of respecting and fostering cultural diversity, nurturing cross-cultural understanding and harmony, mitigating inequalities, and advocating for the rights of marginalized communities. Dumbara artisans' commitment to sustainable practices in maintaining social relations through their craft have enabled them to ensure economic prosperity as well as attain upward social mobility by fostering communal values.

This paper is based on research conducted for my Ph.D. thesis, and I wish to express my gratitude for the scholarship provided by the AHEAD, Ministry of Education, Sri Lanka, for my doctoral studies.

S. De Silva (✉)
Department of Textile and Apparel Engineering, Faculty of Engineering, University of Moratuwa, Moratuwa, Sri Lanka
e-mail: sulari@uom.lk

G. M. Ranathunga (ed.), *Textiles of Sri Lanka*, SDGs and Textiles,
https://doi.org/10.1007/978-981-97-0569-6_11

Keywords Sustainability · Dumbara weaving · Traditional crafts · Artisanal livelihood · Social relationships

1 Introduction: Traditional Crafts and Sustainable Development

This study focuses on the phenomenon of 'traditional crafts' as an attempt to contribute to the contemporary discourse on 'sustainable development', their perseverance and the role assigned to such crafts in the present era. The craft chosen for that purpose is Dumbara weaving, the oldest textile weaving tradition in Sri Lanka that is said to have originated or prospered in the late medieval period [3, 4]. The artisans find their source of income and meaningful life as 'Dumbara weavers' by being hereditary engaged in this craft-making activity. In view of that, the study emphasizes the sociocultural aspects of the sustainability of Dumbara weaving by referring to the United Nations Sustainable Development Goals (SDGs), which include respecting and promoting cultures, facilitating cultural understanding and peace, avoiding inequalities and promoting rights-seeing as those factors are fundamental to human-centered, inclusive, and equitable development through a craft of this type. It is suggested that the perseverance of Dumbara weaving has been through the embeddedness of the livelihood of the Dumbara artisans within social institutions such as family and kinship, as well as through the non-kin artisan–patron relationship, thereby which contributes to achieving the Sustainable Development Goals (SDGs). For as long as Dumbara artisans remain a marginalized inferior group, the relationship they maintain with those who closely associated with the craft has been a decisive factor in determining their artisanal identity and community viability.

Thus, it is hypothesized in this study that the endurance of the Dumbara weaving relies on a reciprocal relationship between the two parties. (1) The first party comprises the Dumbara artisans' kin community who engaged in the craft-making livelihood. They belong to the caste known as *berava* (drummers' caste) which is one of the lowest in the Sri Lankan caste hierarchy. Drumming (at temples and for rituals) was the caste duty assigned to them, yet they were involved in the making of Dumbara weavings as an extra means of livelihood. (2) The second party consists of elite patrons from various backgrounds who support Dumbara weaving: religious and lay regulars, craft retailers and curators, researchers, journalists, government and civic bodies. These patrons concurrently are consumers, enthusiasts, connoisseurs, custodians, and guardians of crafts. A historical examination would reveal several groups of such patrons at different eras in the past—Sinhalese Kandyan aristocrats who provided patronage to Dumbara weaving consuming those products during the era of the Kandyan Kingdom, the English colonialists during the period of British rule, and the owners of handcraft brands and retail outlets or craft boutiques based in Colombo and other main cities in Sri Lanka at the present time.

However, the scholarly discourse on such traditional vernacular crafts is that, the influence and interference of intermediaries into such crafts, including elite patronage, result in diminishing its 'spirit', agency and authenticity [2, 3, 6, 16, 18, 19] (Sinopoli 1998; Scrace 2003, Lass et al. 2009; Sharma 2009). Many instances where traditional crafts have been investigated to construct the aforementioned discourse reveal an unique relationship between artisans and patrons under the conditions specific to each craft, or a generalisation of such relationships across all crafts. Some scholars counter this prevailing discourse by arguing that it unfairly condemns all manifestations of craft patronage, emphasizing its inherent unequal power relationship. They further contend that our understanding of commissioned or sponsored crafts, and the relationship between artisans and patrons, has been skewed by models derived from societies with rigid social hierarchies [5]. Rather than examining the political and economic ideology associated with craft within this discourse, my study aims to understand networks, relationships, niches, and spaces of production and exchange and, to explore how artisans use kin and non-kin relationships to access resources, support, and reciprocity with structural conditions and with forms of change and continuity [19]. Based on Dumbara weaving, this study hypothesizes that the exchange and circulation of such crafts help to build and maintain internal and interpolitical alliances, as well as to reinforce and, in some cases, resist social divisions and hierarchies. The sustainability of Dumbara weaving is ensured by such egalitarian power relations and social negotiations. As far as the particular form of craft in question is concerned, the notion of 'traditional crafts and sustainability', thus, deserves a broader and alternative interpretation and, in turn, has to be considered in the perspectives of the artisans themselves. Accordingly, the study focuses on the ways in which kinship and the artisan–patron non-kin relationships are percieved and accomplished by the Dumbara artisans in order to assure their artisanal identity and communal viability and then to demonstrate how the craft's sustainability is thereby safeguarded.

A primary condition applicable to the selection of a research area for this study was to address how that area contributes to achieve (one or more) Sustainable Development Goals established by the United Nations. Accordingly, the goal sixteenth (SDG 16) seemed to be the most applicable one in relation to Dumbara weaving, especially when considering the Dumbara artisans' community which is oppressed and marginalized. This goal stipulates the promotion of peaceful and inclusive societies, providing access to justice for all and building effective, accountable, and inclusive institutions at all levels. As per this goal, people everywhere should be free of fear from all forms of discriminations, cruelty, violence, and feel safe as they go about their lives whatever their ethnicity, faith, or sexual orientation.

The knowledge and understanding I acquired from over fourteen years of periodic visits to the village of Talagune, coupled with an extensive nine-month stay there, have underpinned the current investigation into Dumbara weaving. By directly engaging with the artisans, I gained insights that allowed me to connect theoretical frameworks concerning traditional crafts with the lived experiences of those who practice the craft.

2 Dumbara Weaving

It becomes clear that some unique inherent characteristics of Dumbara cloths have been crucial for the Dumbara artisans to sustain their craft over generations. Indeed, Dumbara cloths are not consumed by the artisans themselves but is made for the sole purpose of selling. Unlike other crafts, sustaining handloom weaving as a means of livelihood is extremely challenging, especially nowadays, as innumerable substitutes are available for any type of cloth. Dumbara cloths have potential to confront with such situations due to its distinct characteristics which employ artisans' potency and dexterity. In order to gain elite endorsement and therefore become prestige or high-value luxury good, these cloths must possess rare physical characteristics. Anyone who consumes and patronizes a craft that is intricately hand made in this way becomes a beneficiary of multiple factors related to the craft goods as well as its craft tradition. Dumbara weaving therefore can be regarded as "social objects that assume an importance beyond production and utilization. They signify and legitimize group membership and social roles, and become reserves of wealth, storing intrinsically valuable materials as well as the labor invested in their making" [5]. It is therefore intended to devote a considerable portion of this study to describe what Dumbara weaving is, and why it is unique.

Dumbara cloth is of an ornamental weave which consists with intricate patterns and textures. The particular weaving techniques, raw materials (yarn and dyes), apparatuses, and equipment as well as ornamentations in Dumbara weaving are common among other weaving traditions in the world. However, the uniqueness in the use of raw materials, the blending of individual motifs into scattered or striped patterns, the nature of the woven textures with an irregular tactile appearance and touch, and the way in which all the above elements are composed on the cloth enable one to easily distinguish Dumbara weave from any other type of cloth. It is worth mentioning that such unique stylistic features are common to (m)any handmade product.

Even though traditional techniques and processes are employed in making Dumbara weaving, at present the artisans tend to use the ordinary handloom instead of the traditional pit loom (*aluva*, the *al-wale yanthre*—Image 2). As its name suggests, *Aluva* is a weaving apparatus erected on a gutter-like pit. Even though it can be considered a 'machine', it truly is an impermanent device made of few wooden poles, rods, and pieces of ropes and cords. A set of vertical warp threads is stretched between two horizontal poles to prepare for weaving on the loom. One pole (warp beam) is tied to a piece of wooden rod fixed to the ground at the far end, and the other heavier wooden pole (cloth beam) is fixed to the ground on either side toward the front of the weaver. Only this cloth beam and the reed are permanently made in this loom. The loom can be set up at any place and can be removed and folded to be used again. The hut known as *al-ge* where *aluva* is assembled is used for the purpose of pounding and storing sacks of paddy during paddy harvesting times. Hence, weaving activities are not conducted at such times and it is common to see the *aluva* hung on the joist of the roof of the hut. The nature of this loom makes it more compatible with daily activities and household chores of the weaver. According to the veteran artisan Ashoke Padmasiri, the oldest *aluva* in his house belonging to his grandfather was taken to the Ancient Technology Museum located in Polonnaruwa three years ago. Now, among the Dumbara artisans, only Asoka is able to weave on *aluva*. Ashoka, like all other artisans, weaves with ordinary looms but uses *aluva* only for demonstration purposes. These tools and techniques persist, albeit at a nominal level, valued more for their symbolism than their practicality.

Dumbara weavings are produced not as materials but as on-loom finished products. In the past, such products included male clothes known as *thuppotti*, female clothes known as *paada hela* or *sela*, clothes worn by men for bathing such as *diyakacci*, handkerchiefs, towels, *uramaala* (shawls) as well as various other products such as *ethirili* (bedcovers), *sarasili* (house-decorating clothes), *kanappu lensu* (similar to table clothes), chair covers, *ura* (pillow covers), *sivuru hetta* (textiles for making robes), *sivuru pati*, hats, *bulath pahi* (betel bags), *thuppotti* worn by kings and other elite officials such as *nilame*s, *nilamakkara thuppotti*, *thuppotti* worn by ordinary people, clothes of queens, clothes for flags, curtains, and *viyan* (canopy) (these details were provided by veteran artisan A T KSomawansha). Among these products, the *diyakacci* deserves special attention (Images 5 and 6). It is considered a multi-purpose clothing worn by men in the past to cover the lower part of the body. This clothe comprising of a square-shaped decorated cover on the front side, a long narrow piece of cloth on the back side, as well as a long tiny belt to be worn around the waist is a surprise even to veteran weavers as a special technique is employed to weave it as a single piece of clothing. It is noteworthy that even the modern weaving machines which as complex in nature are not able to weave such a product. A few artisans in this village possess models of *diyakachchi* which have been newly woven following old models and techniques. In a book (Thenuwara 2012) composed with the view of recreating a replica of a very old *diyakachchi* to be exhibited at the South Asian Museum in New Delhi, the specific design textures and structures are described extensively (this replica was made by another veteran artisan in the village of Talagune named Gamini Wickramasinghe). At present, the Dumbara artisans pay more attention to making products such as cushion covers, wall decorations, tablecloths, bags, cases, and clothes.

Geometric in many aspects, the Dumbara weavings mainly consist of designs depicting plants, flowers, and animals. Some of the traditional names of such motifs and patterns in Sinhala include: *mal peththa, heen mal peththa, maha mal peththa, pæhena mal peththa, para mal, dæththa, dæthi mala, kathuru mala, ata-pethi mala, idda mala, piyuma, panaa piyuma, kiribath rataawa, hawadi dagaya, thani pota lanuwa, depota lanuwa, ata pota lanuwa, walalu lagaya, lanu gætaya, diya rælla, iri thondu, pannam kuuru, heen rataawa, maha rataawa, dunu mala, gedi mala, awul heraya, bo kola, bopath, aliyaa* (phant), *sinhayaa* (lion), *gaja sinha, kurulla* (bird), *berunda pakshiya, ethkanda lihiniya, nai puttuwa, hansa puttuwa, chaithya* (stupa), *budu pilima* (Buddhaa statues), *deva rupa* (images of gods) (details were provided by Somawansha). The above-mentioned single motifs are combined in many different ways to form sophisticated and decorative patterns on textiles. Even today, the artisans continue to weave archaic motives and patterns in addition to new designs.

At present, artisans purchase (cotton and on occasion rayon) yarn from the marketplace, though the cotton yarn in the past was prepared by the artisans themselves by processing cultivated cotton in the village and the vicinity (cotton was even obtained from areas such as Madugalla). Those cotton yarn as well as the textiles woven using them were known as '*taduppu nool*' and '*tapuddpu redi*', respectively. The *thaduppu* yarn was thicker and raw than the present-day cotton yarn and it is a single strand of yarn. Natural dyeing resources such as the roots, flowers, crusts, fruits, seeds, and minerals were used to color yarn. Even though the Dumbara weavings were complicated in nature, consisting of sophisticated patterns, its color range was limited one to four colors. At present, the artisans tend to make their weavings using yarns of any color that can be bought from the marketplace.

At this point, it would be appropriate to outline some other peculiar features and techniques of Dumbara weaving.

- Dumbara weaving can be considered a method employing a continuous extra-weft technique (weft is the crosswise threads that are passed over and under the lengthwise warp threads) which means that the whole clothe is woven with decorations using a single cross-yarn. A ground weft is added along with each supplementary weft. It is also common to note that a single motif or several such similar motifs placed in a row (especially, images of animals) is/are designed using a method known as '*bema*'. However, it should also be noted that it is quite different to the ordinary '*mal bema*' (according to the method of Sri Lanka), '*jamdaani*', or tapestry weaving methods.
- The design process in Dumbara weaving is also of a special nature. As mentioned above, a colored supplementary yarn is added using a wooden lath (Image 3) to the plain ground to weave consisting of: (1) a single figure in the center of the fabric, (2) rows of patterns running across the fabric, or (3) overall surface ornamentation. In the past, the ground fabric was always woven with an unprocessed undyed raw yarn. The following is a brief account from Dayawathi who is one female veteran artisan in Talagune on the specialty of this technique:

  > Now.....in other weaving types the design is created through the foot treadles. When the number of foot treadles are increased... say by four, eight or sixteen...the number of patterns that can be woven increases as well. The pattern is woven according to the ways that foot treads are pushed. In those weaving methods, the wefts are inserted by raising and lowering the warps using the foot treadles. But, Dumbara patterning are not like that. There are two-foot treadles and two Heddles shafts or harnesses. Here the design is made using a lath which is a narrow smooth piece of wood or bamboo. The lath can raise and lower the warp as many ways as you like and a weft would be inserted (Image 3). This is done...I mean raising and lowering the warp...by hand, but not by feet. Therefore, any kind of design can be woven using this technique.

- The pattern yarn is thicker than the ground yarn used to weave the background cloth since the ground cloth is woven using two strands of yarn while the design is woven with four strands. Hence, it results in an embossed pattern raised against the background. Once woven, it looks like an embroidered work on the cloth.
- In Dumbara weaving, the design appears on both sides of the cloth, making both sides suitable to be used. Accordingly, the negative or a total inversion pattern

of the face side pattern is woven on the reverse side of the cloth. Hence, it can be argued that the Dumbara weaving tradition is one of the most complicated and advanced weaving techniques. If the patterned wefts are not inserted over and under at least for each four to six sets of wefts, it would result in defects such as yarn puckering, ridging, wrinkling, and corrugating. These artisans plan the design by cross-calculating hundreds of yarn strands to compose the patterns proportionally as well as symmetrically. Tangling of yarns, knotting, and planning the edges of the motifs are an intricacy of work that requires a very high level of artistry. All in all, these artisans have already come up with successful problem-solving in relation to the finishing of the highly ornamental woven piece.
- If the design covers the whole piece of cloth, it extends symmetrically running both warp-wise and weft-wise. Initially, the first half of the design is woven inserting spikes such as coconut or palm eakles in between the heald shafts and the warp beam, and then the remaining part is woven effortlessly by repeating and mirroring the halfway design by removing eakles one by one (Image 4).

Why does an inferior minority community weave such ornate cloths that they themselves do not use? How has their weaving qualified to be a tradition or to be called such? How is this community organized for this pursuit? Answers to these questions represent the sustainability aspects of Dumbara weaving.

3 (Re)organization of Production Relations

In this study, I contend that the sustainability of Dumbara weaving is significantly influenced by the reorganization of its production relations, particularly during period of transition (from premodern feudalistic system to capitalist industrial system). I posit that the ongoing process of organizing—or reorganizing—production relations dates back to the 1980s. Before this, Dumbara weaving operated within premodern economic units that were reliant on broader social structures. Craft production that could have been participated by anyone in the kin community in the past was by this time centered on nuclear family. Collective weaving done with relatives was replaced by waged labor, and separate domestic workshops or manufacturing units were established. Each family produces cloths for a particular patron for over decades. At first glance, this way of reorganization of production relations seems destructive; it is true that it leads to competition rather than cooperation among artisans and their families. But, as some scholars described similar cases where artisans carried their kinship relations and their mutual traditions into the new world of work, adapted them in response to the pressures of industrial work, and addressed the industrial system on its own terms (Hareven 2002).

In the past, the production relations of such crafts were organized according to the caste. By this means, Dumbara artisans belong to *berava* or drummers' caste. Though, one or a few members of some families belonging to *berava* caste living in Talagune engaged in drumming as their main livelihood, while another member of the same family wove cloth. Or else, one family was specialized in drumming and sorcery work and another family dedicated to weaving with the participation of all members of that family. Choosing to weave Dumbara cloths is a matter of personal choice rather than *rajakaari* obligation, forming a community of artisans who have opted for weaving as their livelihood, thus transcending the confines of the caste system.

In the past as in the present, there were only a few weaving families or individual artisans lived in this village. About a century ago, there had been three weaving family lineages, namely, *Yakdessa Gedara, Pussekumbure Gedara*, and *Wagalawatthage* (Lawrie 1896) [10]. At present, there are nine weaving families in total, in Talagune, and are descending only from the family of *Yakdessalage Gedera lineages*. A kinship diagram (a genealogical grid) was formulated to reflect the genealogical relationships among the families, which is given in Fig. 1. As indicated in the diagram, those two families belong to the third generation as marked as FAMILY 1 and FAMILY 2. The artisans descend from their great-grandfather Yakdessalage Gedara Punchikira (belongings to the first generation as depicted in the Kinship diagram). FAMILY 1 descends from the children of Punchikira's eldest daughter (Y. G. Tikirathi), while FAMILY 2 descends from the children of his youngest son (Y. G. Padmasiri). It is also evident that the livelihood of weaving is transferred from the elder generation to the following generation depending on the fact that the majority of members of one family have become weavers. These families with a majority of members engaged in weaving represent a form of dominance in the Dumbara weaving (as FAMILY 1 now holds its dominance), while families with fewer members doing weaving tend to gradually give up that livelihood (FAMILY 2). Almost all members of the families who have given up weaving livelihood in this way have already migrated to other geographical areas. Even though Punchikira's siblings were reputed artisans in the past (three of them, one after another, worked at the Kandyan Art Association in Kandy), it seems that their children and grandchildren have abandoned weaving some time ago. Two grand-daughters of Punchikira's one of the younger brothers who resides in the village are still involving in weaving. They each have a loom in their houses. However, they do not work as 'master weavers' and play the role of co-weavers who only carry out the weaving entrusted to them by a relative artisan.

The number of the master weavers in those artisan families currently living in Talagune are eleven and three of them are females. They represent the last three generations of the Genealogical Grid (indicated in dark color) and is a descendant of the Dumbara artisans' lineage. S/he plays a central role in both industrial and family organizations related to Dumbara weaving. A master weaver is a renowned artisan and possesses a high skill in weaving, engaged in weaving on a full-time basis. Such an artisan is an expert in the entire process of weaving ranging from sourcing and processing yarn, fabricating required machinery and tools, planning designs, weaving, and selling the final products. At present, the artisans' households have become small manufacturing units. He/she owns a weaving workshop adjacent to the household and possesses more than one loom as well as other weaving equipment. Such an artisan produces cloths by utilizing wage labor and usually has more than

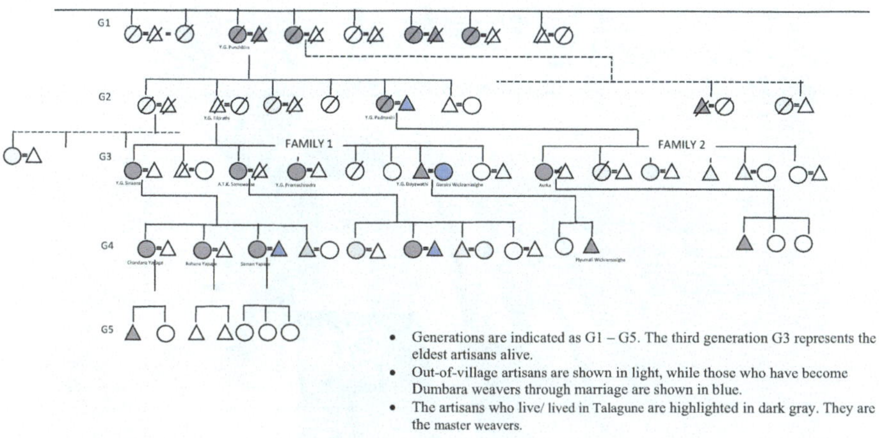

Fig. 1 Kinship diagram; a five-generation genealogical grid of Dumbara artisans

Image 1 Picturesque view of Talagune village; the settlement site of the Dumbara artisan community nestled in the Ududumbara mountain valley in the Central Hills of Sri Lanka [*Photo credit* Author (Sulari De Silva)]

Image 2 Traditional 'aluva' pit loom now used solely for exhibition purposes [*Photo credit* Author (Sulari De Silva)]

Image 3 A Dumbara artisan weaves a traditional pattern using a wooden lath [*Photo credit* Author (Sulari De Silva)]

two co-weavers (assistant weavers) working on a piece rate system. Seasonal co-workers are hired as and when requirements arise especially when they receive orders for a large number of products. The co-workers do only weave using cross-threads (inserting wefts). They not involve in any other task in the weaving process at all. Co-workers mostly come from other villages and train under the master weavers to weave patterns.

Image 4 Coconut sticks are inserted behind the handloom reed to achieve precise mirroring in the design [*Photo credit* Author (Sulari De Silva)]

The artisans' households as work units (mostly as long as the children are unmarried) are mostly egalitarian. A typical household might include master weaver and his/her spouse, their children, and parents and rarely unmarried siblings (mostly if the home is inherited from parents of the master weaver). If the unmarried siblings residing with the family, they would be doing weaving rather than undertaking household chores. The master weaver is the head of the household as well as the weaving workshop. S/he (or the whole family collectively) separates the work process into a number of tasks to perform by a family member. Females wind yarn and males weave cloths (warping is always done by the master weaver). If the master weaver is a female, these tasks are overturned. It had no clearly defined gender division of labor, but this way typically remained unaltered throughout the previous generations. Thus, factors other than technology influenced the organization of weaving, and household differences can be seen in the way work is organized within the family unit. This is the most accepted way of division of labor in Dumbara weaving which occurs voluntarily. The wife of the master weaver keeps busy through other household tasks. She treats their co-workers and visitants with hospitality as it pertains to their

Image 5 Century-Old *Diyacacci* at the National Museum of Colombo, adorned with intricate Rope Knot and Coil Pattern (*Photo credit* Photography Division, National Museum, Colombo)

Image 6 Another *Diyakacci*, showcasing exquisite craftsmanship, on display at the Colombo National Museum (*Photo credit* Photography Division, National Museum, Colombo)

economic deeds. In addition, she stitches the final products (cushion covers, wall hangings, table mats, runners, etc.) from woven material. Children help with any of these tasks, but they are rarely involved. They are always encouraged for schooling. But if the children weave voluntarily, it is also not disallowed. When more orders are received, children are asked to attend to work/support, whether they are trained or not. In this way, over time, the children of the family learn to weave intuitively. Working in such situation is the only way of training as a Dumbara weaver; otherwise, they never receive formal training as apprentices. Another point is that, any one of the family can be involved in economic decisions. In any case, such decisions are made collectively within the family. This nature is common to every artisan's household and running a household this way thus is a communal practice. The focus on household structure and hereditary practices obscures the fact that the family is an economic as well as a moral unit, and members of the household share the labor, hold economic activities, and develop a sense of familial solidarity.

Dumbara weaving now operates, as mentioned, targeting an urban niche market. A master weaver either works for a patron for a long period of time but rarely sells goods in the market to the end-customer. Like the aristocratic landlords of the Kandyan period and the British bureaucrats of the colonial period, the urban upper-class consumers who emerged in the present era demanded craft goods. Artisans like Ashoka accept orders from multiple patrons, while Somawansha supplies his products to only one patron for about forty years.

4 Marginalization and Social Exclusion in Artisanal Weaving Within Pre-modern Societies

Dumbara weaving in the village of Talagune operates in a different way from other few handloom villages in Sri Lanka; it is a livelihood of a kinship-based community established as a social system rooted in the distant past, at least since the era of the Kandyan Kingdom (1597–1815). And it is a craft that has been patronized by the elite since the past. Thence, there are several aspects that deserve attention concerning a craft that originated and sustained in such a context. (1) Dumbara weaving is still practiced in the context of the embedded past in the feudal Kandyan social formation. Prominent among them is the inferior social status ascribed to the artisanal castes (or service castes). (2) However, this geopolitical context has not been supportive in the origination of the Dumbara weaving, nor its long-term existence. Making these pattern weaves known as Dumbara weaving was not part of their caste duties, but rather, as this study suggests, Dumbara artisans themselves maintained this particular type of weaving as a 'reactive attempt' against their low social status. (3) The artisans, through trading in Dumbara cloths, have been maintaining a close relationship with their elite patrons who are on the opposite polar of the social hierarchical system in comparison to theirs. The artisans make these fabulously woven products that suit the aspirations of the elite, though they are deprived of using them.

As mentioned, Dumbara weaving evolved through three main historical eras—the feudal era of the Kandyan Kingdom, the British colonial era, and the post-independence era under a Sri Lankan government. Simultaneously, there are a minority of artisan communities in the lacquer, bronze, pottery, reed mat weaving, and other industries that have made these their livelihood in the hill country region. Even though all these craft industries have come down the years through the same sociological landscapes, Dumbara weaving had a somewhat unique trajectory over the centuries. How it happened will be explained next.

Coomaraswamy mentions that there were two types of textile weavers in Sri Lanka during the Kandyan Kingdom. The two groups are the *Salagama* caste weavers, who migrated from South India and weave fine cloths and *kawanis*, and the local weavers of the *Berava* caste. The local weavers wove unadorned ordinary cloth from the cotton spun in this county, without the influence of the changing fashions of the royal court or the influence of the Indian weavers. Their annual 'tenant' was a loincloth, scarf, or similar cloth. It is called *Alge rajakariya*—duty of loom. These weavers are obliged to perform other services like drumming and sorcery besides weaving. Coomaraswamy mentions that the main duty of this group is weaving, but drumming and sorcery are additional tasks performed by them. Yet, according to John Davy, it is an accepted fact that only a few families living in a certain area of the *Berava* caste are belonging to this community (Davy 1969), which was a more extensive caste in the caste hierarchy in Sri Lanka, lived with textile weaving as their main livelihood. Meanwhile, only the weaver artisans of Talagune have mastered this style of pattern weaving. Caste was an institution that played a significant role in regulating the political, economic, social, and cultural spheres that were formally institutionalized during the Kandyan Kingdom. In the feudal caste stratification, each of these artisan castes was considered inferior to the cultivator caste and represents one of the bottom most strata in the Sinhalese caste hierarchy. These castes primarily provided goods or services to the peasants of the cultivator and artisan castes. The *Beravas* who live in Talagune is a community of drummers, but several families there have been weaving cotton textiles. They have owned very small plots of land (*Nilapangu*) for habitation and cultivation, a few of which are still cultivated today. Evidence is found that these weavers cultivated cotton in small amount in this village (Lawrie 1896, Knox; 1966, Davy 1969) [4] and that the cotton grown in the surrounding villages (*Madugalla, Panaawitiya*, etc.,) was brought to them (Rajapakse 2008). In other parts of the island, the *Beravas* lost their weaving, but to this day, this small community in Talagune now known as Dumabra artisans is still engaged in cotton weaving. As the Kandyan aristocracy (lay and clergy) sought the services of highly skilled craftsmen from other lower castes, Dumbara artisans also received orders by them for their elaborately patterned woven pieces. But none of them ever became a royal weaver the very popular impression on Dumbara weaving so is a misnomer. Some of the families of *Berava* caste Nakati who participated in the drumming service and rituals (*Kankaari shanthi karma*) were recruited by the State Ministry/bureau *Kottal Badda*, and thus those families belonged to a higher stratum of the *Berava* caste (*Maha Badda*). But there was no such role attached to the service of the Royal palace for textile weaving. It was not known that the said bureau carried out activities related to textile weaving or that a separate '*badda*' (a state bureau) was established for weaving and the services of these artisans were attached to it. The King, courtiers, and the Kandyan aristocrats regularly consumed imported clothing and textiles. But it is evident in the examination of the textile pieces that have survived in the ancient period that these decorative textile goods made of Dumbara weaving may have been used for other non-clothing needs of the Royal Palace and elite residences. Instead, if the common people were forbidden to use such intricately ornamented cloths, then these cloths must have been made for the use of the elite. There is no evidence that the weavers' skills and weaving for the elite earned them any privileges such as caste promotion, land grants, or honorific titles. Apart from the *Aachari* caste (smiths of gold and other metals), it seems that the artisans of other service castes made magnificent items did not have the opportunity to have any privileges. What needs to be emphasized here is the inspiration and aspiration of these artisans to maximize their artisanship in such a closed and highly structural social system despite their inferior social status and caste duties remaining unchanged.

Dumbara artisans are also oppressed under the factor of residential segregation. When exploring the demographic factors of this village, it is important to take into account the residential segregation and the spatial distribution of castes as well as *rajakaari* duty. The dwelling area in the village of Talagune which has been assigned as the settlement to Dumbara artisans' community (*berava* caste in particular) in the Kandyan period appears to be the most ecologically disaster-prone area when compared to other adjacent villages and vicinities. Wild animal hazards, landslides, seasonal winds and typhoons, rocky hills, and unfertile soil are major among them. As Lawrie has pointed out in his Gazetteer, I could also understand through my own observations that 'the soil of the Talagune village was not fertile' [Lawrie's Gazetteer of the Central Province (1896: 803)]. Located in a valley surrounded by a ring of mountains, the deepest land plot of this village is the earliest dwelling area of the Dumbara artisans. They did not have an agriculturally self-sufficient economy due to the very few types of trees that grow in this area. Due to this nature of land division and the spatial distribution of castes in the Kandyan region, Talagune can be considered a bi-caste village where the majority *goigama* (cultivators' caste) and minority *berava* caste populations have been living since the past. The *berava* community's position in the Kandyan caste hierarchy is much lower than that of the *goigama* community. Located along the main road from Ududumbara Junction to Kalugala, several villages neighboring Talagune are single-caste villages belonging to the *goigama* caste. All such single-caste villages inhabited by *goigama* communities are served only by Talagune *berava* caste populace. There is evidence to suggest that *berava*s have provided ritual and exorcist services up to areas like Madugalla village. Unlike other villages in the Kandyan region, the lower caste *berava* community in Talagune also had the right to cultivate and, thus, to own limited cultivation land plots as well. As mentioned, there are still a small number of cultivated lands belonging to the Dumbara artisans and some of them have been cultivating paddy. Thus, the *berava* community has to engage in another source of income without depending on cultivation or the natural environs. Some members of the community would have chosen this weaving livelihood as a proper alternative.

5 Crafter to Artisan

It was during the colonial period that this community was progressively able to access social rights, acquire economic resources, and achieve upward social mobility. Their social identity changed from that of a 'crafter' or 'maker of goods' to that of an 'artisan'. The colonial era of Sri Lanka (this article uses the name "Sri Lanka" rather than its former term "Ceylon" from the British colonial period) was the time when the arts and crafts revival movement was growing in Britain, which was the most industrialized nation at that time. Growing anxieties with industrial life during this period led to a positive reevaluation of pre-capitalist production methods and its associated lifestyle. This movement advocated the preservation of traditional crafts, under which the importance of attributing values to the authenticity and agency of the works of art and crafts was emphasized. It was also during this period that new scholarly approaches, such as ethnography and anthropology, spread across various academic disciplines as part of Europe's cultural movement and intellectual reawakening. Accordingly, during British rule in Sri Lanka, the search for information on various aspects of human undertakings had instigated. Although this may contradict the dominant colonial narrative, it is an exploratory process that goes beyond a mere collection of the census and statistical data for the purposes of state administration. The East India Company, the Royal Asiatic Society, the Ceylon Social Reform Society, etc., and the British administrators themselves personally contributed to this mission. Colonial history and the ethnography authenticated the concept of 'craft' and, meanwhile, led to an approach to the preservation of such goods in museums as authentic ethnographic objects. The intricate ornamental goods made by the peasants were also promoted to retailing under the patronage of the British government [9]. In 1882, The Kandyan Art Association was instituted as the first emporium in Sri Lanka aiming at revitalization of traditional Kandyan 'crafts' and provided opportunity for the peasant artisans to sell their products. The Kandyan Art Association was founded by a few British officials who were in charge of the administrative affairs of the Kandy district and maintained it with the financial contributions received from the members of the association. Under this, several veterans from the families of Dumbara weavers were recruited from time to time in the service of this institution. There is an important point to be emphasized about the British government's involvement in the crafts sector in Sri Lanka, particularly in textile weaving. At a time when India's handloom industry was severely suppressed under the British rule, instead of adopting such a repressive policy toward textile weaving in Sri Lanka, it was followed by favoring and providing benefaction. This was because Sri Lanka was not a significant market for Britain's industrially produced textiles, and conversely, handmade textiles in Sri Lanka were not a competitive substitute for them. The British considered exquisite ornamental goods such as Dambara weaving to belong to a special category named 'craft' and 'artifacts' and took measures to safeguard them. This was another dimension of colonial political and economic policy, i.e., these goods were seen as a resource mobilization of colonialism. In the meantime, the Kandyan Art Association made arrangements to publicizing such goods by exhibiting and selling them at trade fairs held in Sri Lanka and across Europe. On such occasions, competitions were organized, and prizes and certifications were awarded for the winning goods. But it was awarded in the name of the Kandyan Arts Association and the artisan did not get any recognition from it. Regardless of its ultimate intention, it is worth mentioning that the measures adopted by British rule toward these goods, used in everyday rural life, enabled them to gain lasting fame and perpetuity. Coomaraswamy's opinion is that due to the abolition of the *rajakariya* system and the lack of patronage of the Kandyan elite, these crafts could have died out during the colonial period if it was not for the Kandyan Art Association [3].

The debate surrounding craft and its mode of production often presents a dichotomy: were artisans merely subjects subordinate to British rule, or were they autonomous agents shaping their production? Additionally, did craft traditions contribute to a global appetite for authentic artisanal goods, thus perpetuating the discourse around craft as marketable commodities, or was "craft" merely an abstract ideal? British colonization in Sri Lanka not only laid the groundwork for capitalism but also fostered a conducive environment for the flourishing of arts and crafts. Consequently, it shouldn't solely be regarded as a catalyst for the decline of peasant agriculture and commodity production, a trend more prevalent in other South Asian nations like India. As for the crafts in the Kandyan region, including Dumbara weaving, it seems that the colonial project has laid a strong foundation for manual workers to become artisans and for products to be artifacts that contribute to the perpetuation of the craft tradition capable of creating a global trend and securing their survival in the capitalist world. This phenomenon radically enhanced knowledge of craft history and aesthetics through extensive collections of artifacts in museums and manuscripts. Through these crafts, the British legitimized and validated Kandyan Sinhalese material culture within the British imperial elitist value regime consisting of foreign bureaucrats, consumers, connoisseurs, and intellectuals as well as for the local elite; even though it was piecemeal and selective, the colonizers' reappraisal of local artistic traditions endorsed an area around which a nascent nationalist cultural identity could develop.

In the late-nineteenth century, during the latter period of the British rule, the nativistic Sinhala Buddhist reformist movement, which developed as an anti-colonial project, seemed to be aimed not only at boosting Sinhalese patriotism and Buddhist religious faith, but also expanding social services in economic aspects by establishing local industries. The main industries focused on were handloom, carpentry, ceramic, and paper industries. At that time handloom weaving was not a major industry in Sri Lanka and it was expected to start textile manufacturing and make it a livelihood for the common people. In 1912, the first textile training school in the country was initiated by Anagarika Dharmapala, who played a pioneering role in this reformist movement, with his family's funding. No doubt the influence of the Khadi movement in India, which was initiated by Gandhi, influenced this. It should be mentioned that unlike in India, in Sri Lanka, apart from agriculture, there was no other manufacturing industry developed at that time. However, Sri Lanka's industrialization movement was not intended to have a broad and vital purpose, such as boycotting imported cloths, or protecting the country's traditional weaving industry, as was the case with the Khadi movement. Khadi movement was a national movement to improve the living of all Indian ethnic groups, but in Sri Lanka it targeted only the Sinhala Buddhists; some castes or races were treated as 'other'. This was not uniformly embraced across the country and no attention was paid to the ancient crafts that were spread throughout the Kandyan area, which was patronized by the British. It should be noted that this movement laid a solid foundation for the expansion and sustainability of handloom manufacturing and other cottage and small-scale industries in the country, especially in the lowland areas. Although in decline today, handloom manufacturing became Sri Lanka's main and largest industry after post-independence era. It cannot be said that this had a direct impact on Dumbara weaving, but this overall development in the field of the industrial sector, mainly in handloom manufacturing, has been beneficial in the long run.

Even after gaining independence, the necessity for a Sinhala national movement and ideology remained steadfast, with no apparent ideological shift. Ancient architectural ruins and medieval arts and crafts continued to be revered as cultural symbols, serving as tangible evidence of Sinhalese artistic prowess and reinforcing Sinhalese chauvinism, even though some were created by non-Sinhala-Buddhists. These objects attained national significance as tangible links to the country's heritage. Through this, traditional crafts were classified under the nomenclature of 'indigenous', 'vernacularism'. This tendency, observed globally in the appreciation of traditional folk arts and crafts, also emerged in post-colonial states across South Asia.

For this reason, intervening in this field through the formulation of institutional settings and policies for the upliftment of arts and crafts had become a primary task of the sovereign nation-state of Sri Lanka. In the post-independence era with the advent of *democracy* and *socialism*, instead of focusing on cultural welfare where the arts and crafts could be used to address social issues and enhance the nonmaterial aspects of public life, the government looked to promote soft power, national image building to the sake of political power. There even economic development was considered inferior. Ironically, by the 1960s Sri Lanka have had a strong arts/crafts infrastructure—state-owned entities and institutes, universities, training institutes, arts programs, companies, galleries, arcades, and festivals—that was surprisingly invisible outside. There was a very large governance structure—departments and institutions under central government ministries, and at the provincial, and village levels. National Craft Council, National Handcrafts Board, National Design Center, Industrial Development Board, Department of Cultural Affairs, Department of Textiles are few of them. Subsequent government administrations conscripted their own national cultural plan, which by the 1960s aimed to promote the excellence of the ancient arts and foster contemporary art. My contention is that although the initiatives and measures were outstanding and strategically important, they were not succeeded under such a large and unsustainable public sector and a risk-averse culture. Sri Lanka's populist–nationalist policies have resulted in many state-owned enterprises becoming dysfunctional and operating at massive losses. I present two case examples related to my field of study on Dumbara weaving. The *Laksala* handicraft outlets' chain owned by the Sri Lanka Handcrafts Board was established during this period (in 1964) with the aim of marketing handcrafts locally and globally. It provides financial and advisory facilities to the craft industry in rural areas through its affiliates. The *Laksala* has become a very loss-making state-owned enterprise for a long time and the Dumbara artisans do not intend to sell their products through *Laksala* because of the strains and losses they have to experience in dealing with such government institutions. Another such project is that the establishment of craft villages and the settlement of artisans in newly formed villages under a program for the upliftment of Kandyan traditional crafts. In 1968, a craft village (*Kalaa Pura*) was established in *Natharampotha* and selected families of the artisan communities who lived in the villages were resettled in this village. The craft village project that was aimed at local and foreign tourism had established a design center and a sales outlet in addition to its domestic workshop units. Although the village is still functioning today, it does not seem to have been a success for Dumbara weaving and few other crafts.

By the late 1960s, the next craft revival has emerged as a commercial trend. The consumption and patronage of handcrafts, especially traditional crafts such as Dumbara weaving, became very popular among the new urban elite in the low country area. This trend then gained mainstream popularity and lasted for decades. The governing authority of the above-mentioned state institutions set up for safeguarding crafts was vested in the higher echelons who were involved in the private craft businesses. Siva Obesekara, who was the founder chairperson of *Laksala*, organized the Sri Lanka Craftsmen's Association (SLCA) in 1964 and served as its president for many years. She is descended from a noble lineage and was later elected as a Member of Parliament and a Minister of the government. To promote the handcrafts of the artisans of SLCA, she simultaneously set up a handcraft emporium called *Lakpahana*. Siva was also the president of the World Crafts Council. No matter how much these artisans benefited from such initiatives, and it is the dominant elitists who recapture the social honor, privilege, and financial surpluses even within these new value regimes. Even today in Sri Lanka, all these state-owned institutions are governed by private industrialists in the respective sectors. Politicians use rural craft industries for their private businesses through government projects, such as rural industrial development and women empowerment, and otherwise interfere with them (under the previous government, the spouses of the President, Finance Minister, State Ministers, Members of Parliament, Provincial Governors, etc.). Contemporary Sri Lankan craft is a signifier of indigenization, Sinhalese artistry, noble ancient tradition, and craft-based aesthetics. However, it has led to revitalization centered on concerns about artisanal industries, markets, and consumer identities. In any kind of process that is happening at any time, through these unique forms of goods, the elite status is consolidated and the need for sustainable elite intervention is justified. Crafts verify the status of the artisan in these regimes.

6 Livelihood Improvement and Upward Social Mobility of the Dumbara Artisan Community

By the 1980s, the situation was changing and it was the period when Dumbara weaving faced the most transformation. In the early 1980s, a few leading privately owned art and craft conglomerates that were very successful in manufacturing and retailing of handcrafts came forward to invest in traditional crafts. Since then, Dumbara weaving has become a patronage-based industry. Dumbara artisans, some of whom work for only one patron, while others carry on weaving on commission from several other patrons. They made their 'artisan made' mark or the 'brand identity' through the patrons' commissioned works. Artisans were driven by the sheer necessity to survive to become increasingly entrepreneurial by attempting to create market niches. Due to the rapid development of tourism and the new tendency of the Sri Lankan diaspora community to use these items as exotic souvenirs, the value of traditional crafts increased significantly during this era. It can be evidenced that this situation affects Dumbara weaving more lucratively. The influential concepts of yachting in the contemporary world seem to be for a more progressive and for the sake of a better world; thrift, environmental and ethical activism (green/ethical production), anti-consumerism (conscious/responsible consumption), localization, heritage, nostalgia, *objects of virtu*, shared aesthetics concepts.

Contrary to these trends, it can be seen that new economic ventures targeting the village communities were being started during this period, yet, mainly at the domestic level. A handful of populaces who stayed in the village without migrating or returned to the village after quitting their jobs is now engaged in agriculture and petty commodity manufacturing activities based in village. The provision of infrastructural facilities to such sectors under various programs by the government and non-governmental organizations, provision of financial aid and other resources, and implementation of training programs and marketing facilities were increased during the later decades. It was a direct result of this progression that the producing of craft goods in the Kandyan region was transformed into petty artisanal enterprises. Even though rural enterprises are usually carried out with the engagement of intermediates such as external traders and merchants, the products manufactured in these villages are sold directly by the farmers and commodity manufacturers themselves to the end-consumer. In particular, as production is done here using limited human labor and other resources, it does not generate a considerable surplus that allows outside traders to intervene. The gain or profitability of these products is minimal as less is produced. The characteristics of a typical rural economic system of producer–intermediary relations are not visible in the case of Dumbara weaving; rather, it is a patron–artisan relationship that only works for art/craft economies. Apart from the emergence of a bourgeoisie of petty traders in Talagune and other adjacent villages, only the community of Dumbara weavers has become petty commodity manufacturers and prospered to such an extent. In that regard, the weavers' community has now surpassed the economic status of the upper *goigama* caste.

The proliferation of industrial endeavors across the nation facilitated infrastructure advancement, notably the augmentation of power distribution, rural road networks, enhancement of local public services (such as administration, education, and healthcare), and the establishment of small trading centers and administrative hubs surrounding select villages. Due to the expansion of public services and the establishment of government institutions in the regional level, more government employment opportunities were created meanwhile for the village communities. A significant number of people employed in the government sector (nursing, teaching, clerical and security services, banking, etc.) are now permanently settled in in these rural villages. However, only those members of the *goigama* community who had completed formal education were able to obtain permanent upper and lower-middle-class jobs with regular salaries. Under these new circumstances, despite having minimal educational qualifications, at least a few of the descendants of Dumbara artisans were able to be employed in the regional branches of government-owned textile institutions (Kandy, Digana, and Kundasale) during this period. But the employment opportunities open to such low-caste people were limited to low-paid, casual, temporary, and laborer positions. In terms of private sector employment, a significant number of populaces of these villages have been working in large-scale livestock farms in nearby villages. Livestock farming is the sector that has created the majority of employment opportunities for people of all castes. In addition to that, there is another significant number of people who provide labor services (such as carpentry, electrical, and plumbing services) for a daily wage in industrial sectors such as building construction. It is very true that the diversification of income generation in these villages was very limited even during this period, but on the other hand, it must be said that these villagers chose such menial and casual jobs because they wanted to contribute to the cultivation activities of their families. This situation is crucial for the case of Dumbara weaving as well. Some members of the weavers' families who are employed in this type of salaried jobs have not given up their participation in weaving, which is still being carried out at their ancestral home.

In discussing the economic situation and the employment opportunities that had been emerging in these villages since the 1980s, I would like to mention some important facts about the formal educational sector as well. Even before independence, i.e., the free education policy came into effect in 1945 was planned to provide regular, equal, and free education opportunities to every citizen of the country, but due to various socio-economic issues, it was not implemented realistically until the 1980s. Only a few were able to complete school education in the past decades. By the 1980s, the highest percentage of those who were born in the 1970s was increasingly interested in pursuing educational opportunities. It is noteworthy that there were schools especially built for low-caste children in villages, especially in the Kandyan region. Thus, Talagune village also had a small junior school built in the late 1950s. Even though there were attempts to say that village schools were built for children of all levels, children of *goigama* families did never attend them or, on the other hand, there were cases in which *non-goigama* low-caste children gave up school education due to caste discrimination in such schools. However, after the 1980s, children of almost every caste or level were able to attend schools. Interestingly, the second generation of the weavers living in the village of Talagune has gained secondary education not from the school in the village, but from the Ududumbara Maha Vidyalaya located in the closest town. As the parents had changed their caste identities (caste-oriented surnames and personal names), caste-based discrimination was at a minimum level when their children attended schools in the town area. It was due to the fact that even low-caste children of *beravas* do not attend the village school in Talagene, that it had to be closed a few decades ago. Accordingly, very few of the oldest generation of artisans living in this village (the third generation according to the genealogical grid; Kinship diagram—Fig. 1) have received even basic education from the village school, but the next generation (fourth generation in genealogical grid; Kinship diagram—Fig. 1) completed their schooling up to the secondary level from schools both in the village and in the town. The youngest generation goes to school entirely outside the village; they even go to schools in main cities (Digana and Kandy) and have diplomas and higher vocational qualifications as well. This was the decade when the expansion of educational opportunities for all village populaces instigated. Based on the fact that the state education system provides salary employment, it appears that the penetration of monetary relations into the village did not significantly change the socio-economic context of these villages. What was happening further was the migration of educated people out of the village.

The penetration of money into villages has resulted in some progress in these villagers' daily consumption pattern, enabling them to have a better quality of life with more material possessions. The transformation of the Ududumbara town into a small 'bazaar town' has made it possible for them to buy a limited range of consumer goods. Over the past decade, I have witnessed that how the artisans' houses built with *iluk* (a kind of thatch grass) roofs and walls made of mud and sticks (*katu-meti*) are being transformed into houses made of brick and cement with asbestos roofs. Almost every family now owns a kind of single occupancy vehicle—a foot-cycle, a motor bicycle, or occasionally, a three-wheeler or a small lorry—so it seems that there has been a significant change in the motorized modes of transportation in the village as well. Instead of irregular public transport services, this is a more desirable option for the villagers because of its efficiency and affordability.

It was under such circumstances that these weavers have been maintaining this craft industry over centuries in a lively manner by reacting as well as being absorbed into contexts.

7 Patronage and Kin Orientation for Dumbara Weaving

Currently, the domestic system also referred to as the 'putting-out system,' prevails in the crafts sector, wherein peasant artisans labor from their homes while merchant employers provide materials to them. The artisans then return the finished goods to the employers, who compensate them based on piecework or wages. In contrast to this usual economic transaction between mass production and craft trade, or between artisan employees and tradesmen, the interaction between artisans and patrons in

Dambara weaving embodies a reciprocal social bond. The patrons of Dumbara weaving were not just intermediaries or merchants; they symbolized a privileged class deeply involved in art circles, eventually emerging as the driving influence behind the art scene. Their involvement with Dumbara weaving has led to the evolution of a peasant craft into an esteemed art form. It could be stated that such a patronage system demonstrates alignment with Costin and Arlignton's perspective. It implies that Dumbara artisans possess the capacity to move beyond conventional labor roles, nurture social connections, and reshape social structures. Consider the following statement by Costin and Arlington "it is important to see producers as active, multi-dimensional participants in production–distribution systems for yet another reason… Consumers value crafts for their utility, their prestige value, their political significance, and their symbolic, ritual, and/or ideological associations. Recent studies of material culture stress the active role of objects in creating social relationships and transforming social organization. Amidst this interest in the dynamic nature of material culture, we cannot accord a passive role to the creators of that symbolically charged material culture, since it is the artisans who often impart value and meaning to their goods through the creative process" [5]. The correlation or the relationship mentioned does not undermine the autonomy and agency that Dumbara artisans have over their craft.

Artisans maintain a consistently optimistic perspective on their patronage. Quoting from discussions with veteran artisan Sirisena, here's what he expressed:

At a time when there was a lack of work for a long time, I met a gentleman named Samarawikrama. He came in search of me. I met Mr. Samarawickrama in 1989. Since then, my children and I still mostly do the work he commissioned. As he was an architect, he also received many orders for the design of articles for home decoration… I learnt a lot while working with him. Honestly, it wasn't until I collaborated with him that I recognized my potential for more intricate designs. I gained insights on color coordination, preserving the texture of a fabric, and ensuring its quality when introducing it to the market. Now I am much more improved than when I was in the past…..I didn't realize I had such talent until he helped me develop it to this level. Now when I compare my old designs with the ones I made for him, those are at a much lower standard.

In addition to adhering to the traditional patterns and designs of Dumbara weaving, patrons introduced novel designs. Consequently, artisans were compelled to explore fresh ideas. This inspired them to experiment with innovative techniques to fulfill these evolving demands. Artisans develop specialization in particular patterns through the creation of their unique designs over time, leading to a diversification of artistic styles through such a patronage system (Images 7 and 8).

Sirisena: *When creating designs, I had to go to Mr. Samarawickrama's house in Colombo. There I would stay for a couple of days and we would create the design together. He would keep me by his side and draw the design on his computer….*

….I remember well what Mr. Samarawickrama said when he first came here. "I want to do something new. Would you like to take the 'Talagune method' and do something new?" he asked after seeing our old designs. Some of our old motifs and patterns are included in the designs but they are not given prominence.

Image 7 Symmetrical elegance of Somawansa family's signature patterns [*Photo credit* Author (Sulari De Silva)]

Image 8 Master artisan Sirisea's tapestry designs: some of non-traditional novel designs [*Photo credit* Author (Sulari De Silva)]

There is a pattern of ours called the 'Kinnara coil'. It is similar to the squares in a checkers board and are like pebbles in designs he developed. That motif is used often. The 'wave' pattern is also used. He did not want us to use the traditional method but the tapestry technique instead. In our method the cross thread is drawn from one end to the other (from the right selvage to the left selvage of the cloth) and it is the sole thread used throughout the whole cloth. But, it's not the same with the tapestry method.

Below is the view of veteran artist Mr. Somawansa. He has been producing his fabrics exclusively for 'Barefoot' for nearly forty years.

> *The Barefoot Company does not give us designs. What they would do is..............they would choose from the designs that we would take there, to the Barefoot Company .... Picking the samples they would say "this is good....this is good" and choose. Then they give orders and ask us to bring a particular number from the chosen designs. Then we would weave accordingly. From what we have given to Barefoot...the changes have been mostly to colors. They would say, "It's better if this flower comes here, and make it this color", and things like that. Other than that, we don't have to completely change designs and patterns. Then it would lose its value. They preserve the ancient patterns even better than us. They don't try to change the patterns according to their desires. They try to preserve it and take it forward.*

The craft patronage will also affect their community viability as well as improving artisanal potency. Dumbara artisans' community is reciprocal in the present time in that it is concerned with a 'kinship orientation' to the modern world crafts rather than internally correlated kinship systems. This kinship is oriented toward a new craft entrepreneurship that goes beyond past methods of training and mastery, sharing of knowledge and materials, collective trade, etc.

The artisans have to maintain reciprocity between their kin families and have to solely rely on themselves when faced with rejection in the village due to low caste factor. It is important to protest as a community by building alliances with each other against the issues they face, such as low caste discrimination and oppression. I am presenting one example that was told to me. It was only recently that this village received electricity. However, the villages surrounding Talagune had received electricity many years ago. According to the artisans, this was an attempt by the other high-castes to prevent them from attaining prosperity through the Dumbara industry. The weavers state that people of other castes (high-castes) in top government positions worked to prevent their village from receiving electricity. After many failed protests, the weavers had reached out to friendly officials at the Textile Department. These officials had reached out to the central government with weavers' appeal that 'as traditional Sri Lankan artisans they need electricity to work (as they cannot work in the night with the light from kerosene lamps)'. This request for electricity had been given special approval and the village received electricity. During this discussion, Chandana showed me the files that he still possesses in relation to this matter including letters, replies, petitions that were sent to and received from various government institutions. This is one instance where these weavers were able to win equal rights for their community due to the close relationships that they maintain with high-ranking officials.

Additionally, they collectively construct myth of origin of the Dumbara weaving, histories, and traditions to be compatible with elitist ideologies. These aspects are the cultural capital of the artisans. These ideologies and identities are essential for artisans' economy as well as for their upward social mobility. Hence, there is a need for collaboration, cooperation community viability.

8 Conclusion

The contribution made by Dumbara weaving toward the sustainable development is not so much perceptible. In particular, a mere glance at this craft in which the involvement is made by a minority community consisting of a few peasant families (less than 10 families) shows that its contribution to the sustainability is significantly minimal. Dumabra weaving as a traditional or hereditary craft is considered as a part of the history and has impact upon the economy of the region in which they flourish, as well as for the whole nation are essential for social wellbeing, for identity. The sustenance of crafts within communities plays a crucial role in fostering social well-being, shaping identity, and preserving traditional knowledge. This study highlights the significant influence of social relationships, which often surpass economic transactions, in ensuring the longevity of artisan communities and their cultural heritage.

References

1. Appadurai A (1988) Introduction: commodities and the politics of value. The social life of things: commodities in cultural perspective, pp 3–63. Cambridge University Press
2. Black A, Burisch N (ed) (2020) The new politics of the handmade: craft, art and design, Bloomsbury visual arts
3. Coomaraswami AK (1907) Native arts & handicrafts, twentieth century impressions of Ceylon. In: Wright A (ed) Lloyds Great Britain Pub. Co., London
4. Coomaraswami AK (1908) Medieval Sinhalese art. Pantheon Books Inc. New York (1956 ed.)
5. Costin CL, Arlington RPW (1998) Craft and social identity. American Anthropological Association, Archaeological Papers of the American Anthropological Association, No. 8
6. De Silva S (2023) Crafts' value paradox; political economy of Dumbara weaving. Social Scientists' Association (SSA), Colombo
7. Dormer P (1997) The culture of craft: status and future. Manchester University Press
8. Herzfeld M (2004) The body impolitic: artisans and artifice in the global hierarchy of value. University Press, Chicago
9. Jones R (2008) British interventions in the traditional crafts of Ceylon (Sri Lanka), C.1850–1930. J Mod Craft Berg
10. Lawrie AC (1898) A gazetteer of the central province of Ceylon, vol. II. In: Skeen GJA (ed) Government Printer, Ceylon
11. Nash J (ed) (1993) Crafts in the world market: the impact of global exchange on middle American artisans. State University of New York Press, Albany
12. Rathanasara T (1954) Ancient textiles of Ceylon. Ceylon Today 4
13. Risatti H (2007) The theory of crafts: function and aesthetic expression. The University of North Carolina Press
14. Scrase TJ (2003) Precarious production: globalisation and artisan labour in the third world
15. Sharma K (2020) Hand-crafted identities: sartorial taste and belonging amongst elite women in urban India. Sage J 25(1)
16. Sinopoli CM (2003) The political economy of craft production: crafting empire in South India, C.1350–1650. Cambridge University Press
17. Thenuwra C (2012) Handwoven heritage—future foundation: Diyakacci—fabric design analysis, Item 07-97-217, National Museum Colombo

18. Venkatesan S (2009) Rethinking agency: persons and things in the heterotopia of 'traditional Indian craft'. J R Anthropol Inst
19. Wilkinson-Weber C, Denicola AO (2016) Introduction: taking stock of craft in anthropology. In: Critical craft: technology, globalization, and capitalism. Bloomsbury, London

The manufacturer's authorised representative in the EU is Springer Nature Customer Service Centre GmbH, Europaplatz 3, 69115 Heidelberg, Germany. If you have any concerns regarding our products, please contact ProductSafety@springernature.com

Printed and bound by CPI Group (UK) Ltd, Croydon, CR0 4YY
01/10/2026
02225862-0005